IntechOpen

Supporting open minds since 2005

Molecular Medicine
http://dx.doi.org/10.5772/intechopen.78531
Edited by Sinem Nalbantoglu and Hakima Amri

Contributors
María Del Rocío Reyes-Montes, Esperanza Duarte-Escalante, María Guadalupe Frías-De León, Erick Martínez-Herrera, Gustavo Acosta-Altamirano, Xianquan Zhan, Na Li, Xiaohan Zhan, Liming Shen, Sijian Xia, Huajie Zhang, Fang Yao, Xukun Liu, Yuxi Zhao, Ming Ying, Javed Iqbal, Qiong Liu, Munindra Ruwali, Dingan Zhou, Jiawei Zeng, Pingsheng Hu, Xiaodong Su, Xing Zeng, Yadong Li, Zhixiong Wu, Xing Wan, Laure Cayrefourcq, Catherine Alix-Panabières, Sinem Nalbantoglu

First published in London, United Kingdom, 2019 by IntechOpen
IntechOpen is the global imprint of INTECHOPEN LIMITED, registered in England and Wales, registration number: 11086078, 7th floor, 10 Lower Thames Street, London, EC3R 6AF, United Kingdom
Printed in Croatia

British Library Cataloguing-in-Publication Data
A catalogue record for this book is available from the British Library

Additional hard and PDF copies can be obtained from orders@intechopen.com

Molecular Medicine
Edited by Sinem Nalbantoglu and Hakima Amri
p. cm.
Print ISBN 978-1-83962-759-0
Online ISBN 978-1-83962-760-6
eBook (PDF) ISBN 978-1-83962-761-3

For EU product safety concerns:
IN TECH d.o.o., Prolaz Marije Krucifikse Kozulić 3, 51000 Rijeka, Croatia, info@intechopen.com or visit our website at intechopen.com.

Meet the editors

Dr. Sinem Nalbantoglu is senior investigator in the field of molecular medicine and bioengineering, and chief senior researcher at the Laboratory of Molecular Oncology at the Gene Engineering and Biotechnology Institute, TUBITAK Marmara Research Center. She performed her postdoctoral studies at Georgetown University School of Medicine and National Institutes of Health. Dr. Nalbantoglu's research and teaching focus on systems and molecular medicine, including omics-based clinical biomarkers for precision oncology and autoinflammation. She holds a TUBITAK R&D Women Entrepreneur Innovation Award, and works as principal investigator, senior researcher, and co-investigator of several partnering projects; she also supervises MSc and PhD students. She publishes her work in peer-reviewed national and international journals, and has authored/coauthored book chapters and conference proceedings. In addition, she has presented national/international invited talks, and has been a reviewer for many scientific journals.

Dr. Hakima Amri is a professor at Georgetown University School of Medicine in the Department of Biochemistry and Cellular and Molecular Biology and the Division of Integrative Physiology. She holds a Master of Science degree in Reproductive Physiology and a PhD in Steroid Biochemistry from Pierre and Marie Curie University-Sorbonne, Paris, France. She then moved to the United States, where she joined the Georgetown University Faculty. Dr. Amri's research focus is on understanding the action mechanism of disease processes using comprehensive systems biology approaches. She investigates cancer biomarkers and bio-signature identification using big data. Her project is enhanced by the novel analytical approach she recently developed in collaboration with her colleagues. She uses this novel method to translate cancer omics, from genomics, proteomics, and metabolomics high-throughput data, into molecular signatures that are then presented in a multidimensional and dynamic model best suited for precision medicine. Dr. Amri is currently working on translating Phylomics® (http://phylomics.com) into a cost-effective and noninvasive diagnostic tool. Her work is supported by a recently granted patent by the US Patents Office.
Dr. Amri has taught a significant number of students in classroom and laboratory settings, spanning undergraduate, graduate, doctoral, and postdoctoral levels. She received funding from both government and private sectors. Her scholarly activities are marked by her published work in peer-reviewed journals, presentations at national and international conferences, and authorship of a book on Avicenna's medicine, several book chapters, and reviews.

Contents

Preface

Molecular medicine is an applied science focused on human genes/transcripts, proteins, metabolites, and metabolic networks that describe molecular and cellular processes of health and disease onset and progression. Molecular medicine aims to define associated predictive, diagnostic, and treatment assessment biomarkers for personalized diagnostic/therapeutic strategies, and for a targeted approach to providing the best healthcare for the twenty-first century. Here, in *Molecular Medicine*, a collection of scientific contributions addressing specifically the field of molecular biology and genetics, genetic epidemiology, laboratory medicine, molecular diagnosis, next-generation sequencing, mass spectrometry, omics technologies, biomarker discovery, liquid biopsy, cell heterogeneity, circulating tumor cells, cancer, *inflammation,* infectious diseases, systemic and chronic diseases, neurodegenerative disorders, pharmacogenetics and pharmacogenomics, and precision/personalized medicine is presented.

In the first chapter entitled "Introductory Chapter: Insight into the OMICS Technologies and Molecular Medicine" by Sinem Nalbantoglu and Abdullah Karadag, an overview of the latest headlines of molecular medicine, including promising research strategies and their emerging roles in biomedical research, has been compiled. Throughout the chapter, primary objectives of molecular medicine, involving prediction of potential future pathologies, identifying disease state through effective screening and early diagnosis systems, decision-making on effective treatment strategies, monitoring prognosis and health status, as well as predicting recurrence earlier to apply alternative treatments, have been emphasized.

In the chapter "Precision Medicine—Role of Biomarkers in Early Prediction and Diagnosis of Alzheimer's Disease" by Liming Shen, Sijian Xia, Huajie Zhang, Fang Yao, Xukun Liu, Yuxi Zhao, Ming Ying, Javed Iqbal, and Qiong Liu, the authors precisely outline clinical biomarkers and precision medicine for Alzheimer's disease, which is a chronic and irreversible neurodegenerative disorder. In this chapter, the authors introduce promising research efforts and progress in the development and validation of invasive, minimally invasive, and non-invasive personalized diagnostic strategies and their applications, neuroimaging techniques and neurochemical assays, novel ultrasensitive immunoassay, genetic testing, mass spectrometry methods, metabolomics, and exosomes.

Munindra Ruwali, the author of the chapter "Pharmacogenetics and Cancer Treatment: Progress and Prospects," contributes a comprehensive review on the developments and recent advances in cancer research, pharmacogenetics, and oncogenetics. As precisely addressed by the author throughout the chapter, cancer pharmacogenetics has great importance for considering genotypic and phenotypic heterogeneity among patients, and offering personalized treatment leads to improved response to therapy and decreased side effects in individuals.

Laure Cayrefourcq and Catherine Alix-Panabières, in their chapter "CTCs as Liquid Biopsy: Where Are We Now?," comprehensively examine and summarize the latest findings on detection and characterization strategies of circulating tumor cells (CTCs) in different tumor types together with their advantages and challenges. The authors also discuss other potential circulating biomarkers to be used as liquid biopsy in oncology, and conclude that liquid biopsy diagnostics might be a promising oncotheranostics strategy.

In the chapter by Na Li, Xiaohan Zhan, and Xianquan Zhan, entitled "Energy Metabolism Heterogeneity-based Molecular Biomarkers for Ovarian Cancer," the authors search for promising clinical biomarkers for ovarian cancers considering the fact that energy metabolism heterogeneity, namely the Warburg and reverse Warburg effect, coexist in cancers. As thoroughly explained by the authors, the uniqueness of their approach is to look beyond blocking both the Warburg effect and the reverse Warburg effect. Indeed, by implementing an integrative analysis of transcriptomics, proteomics, and mitochondrial proteomics, the authors provide new molecular insights into the promising energy metabolism-based target treatments for ovarian cancer patients.

In the chapter "A novel P53/POMC/Gas/SASH1 Autoregulatory Feedback Loop Activates Mutated SASH1 to Cause Pathologic Hyperpigmentation" by Ding'an Zhou, Jiawei Zeng, Pingshen Hu, and Xiaodong Su, the authors focus on the interaction of p53 with other transduction pathways in a cell model. Their approach is directed to enrich and redefine the p53-responsive genes and their associations. The findings led to the conclusion that understanding the interactions of the p53-responsive genes will elucidate the p53-programmed responses to stress and pathological conditions.

The chapter entitled "Molecular Diagnosis of Invasive Aspergillosis (IA)" by María del Rocío Reyes-Montes, Esperanza Duarte-Escalante María Guadalupe Frías-De-León, Erick Obed Martínez-Herrera, and Gustavo Acosta-Altamirano refers to the epidemiology and molecular diagnosis methods for the detection of the disease and associated challenges. As explained by the authors, molecular diagnosis is superior to the conventional tests in terms of detection reliability, sensitivity, and specificity.

The chapter entitled "Metabolomics: Basic Principles and Applications" by Sinem Nalbantoglu examines the latest and most promising of the emerging high-throughput omics technologies, as one of the system's medicine components, metabolomics. The chapter offers a brief overview of the field focusing on methodological advances, strategies, and challenges, as well as current and future bioapplications.

Principally, *Molecular Medicine* provides a glimpse into the latest developments of systems and molecular medicine, highlighting the emerging high-throughput technologies, promising potential applications, and progress in the development of technological and biomedical strategies. The book offers supportive updated content for both medical students and healthcare providers.

Dr. Sinem Nalbantoglu, PhD
Molecular Oncology Laboratory,
Gene Engineering and Biotechnology Institute,
TUBITAK Marmara Research Center,
Kocaeli, Turkey

Dr. Hakima Amri, PhD
Professor of Physiology, Biochemistry, Graduate Education,
Department of Biochemistry and Cellular and Molecular Biology,
Division of Integrative Physiology,
Georgetown University Medical Center,
Washington, DC, USA

Chapter 1

Introductory Chapter: Insight into the OMICS Technologies and Molecular Medicine

Sinem Nalbantoglu and Abdullah Karadag

1. Introduction

Molecular medicine aims to reveal molecules, such as genes, transcripts, proteins, and metabolites, to underlie the mechanism behind the physicological processes as well as alterations during the pathological conditions at the cellular level. Furthermore, molecular medicine intends to improve public healthcare and disease management through development of biomarker-based screening, diagnostic, and monitoring systems as well as target- and mechanism-based treatment strategies. Human Genome Project has been completed in 2003, exactly 50 years after Watson and Crick invented DNA structure. Based on this valuable breakthrough, the twenty-first century's molecular medicine approaches have been attributed to identify and understand functions and interactions of human genes to shed further light on health and disease mechanisms at the basic molecular and cellular level.

Published by James Watson in the first edition of "The Molecular Biology of the Gene" (1965), the central dogma of molecular biology was a complete demonstration of the flow of genetic information basically described as DNA makes RNA, which in turn makes proteins: DNA → RNA → protein [1]. However, later on the 1980s–1990s by applying improved molecular biology methods, the single gene and inheritance concept has changed to multiple genes and inheritance with interactions of genes, RNAs, proteins, and environment in a particular cell. Extinction of central dogma has led to proper and critical understanding of diseases and generation of molecular medicine. By this way, a new concept called phenome, as the total phenotypic characteristics of an organism, has emerged, which implies interaction of the whole genome with the environment [2].

Primary objectives of molecular medicine includes predicting potential future pathologies, identifying disease state through effective screening and early diagnosis systems, decision on effective treatment strategies, monitoring the prognosis and health care, and predicting recurrence earlier to apply alternative treatments. In this regard, molecular medicine aims to obtain decreased under/over/mis-diagnosis and generate effective targeted therapies without side effects. Here, we provide an overview of the latest headings of molecular medicine including promising research strategies and their emerging roles in biomedical research.

2. OMICS technology

The terms "Ome" derived from a Greek word and "Omics" are derivations of the suffix -ome which means "whole," "all," or "complete." With the addition of -ome to cellular molecules, such as gene, transcript, protein, metabolite, it can be referred as genome, transcriptome, proteome, metabolome, respectively [3, 4].

Omics technologies and systems biology are the emerging concept of molecular medicine (**Figure 1**). Omics refers to collective and high-throughput analyses including genomics, transcriptomics, proteomics, and metabolomics/lipidomics that integrated through robust systems biology, bioinformatics, and computational tools to study the mechanism, interaction, and function of cell populations' tissues, organs, and the whole organism at the molecular level in a non-targeted and non-biased manner [5].

Genomics is the systematic study of an organism's entire genome [6]. The human genome is made up of DNA (deoxyribonucleic acid) comprising approximately 3 billion base pairs of four chemical structures (adenine, guanine, cytosine, and thymine), also called nucleotides. DNA contains genetic information required to build and maintain cells. A gene denotes a specific unit of DNA that hold information to make a specific functional unit named protein. It is estimated that the entire human genome contains approximately 21,500 genes. The order of the nucleotides reveals the meaning of the information encoded in DNA. Emergence of high-throughput sequencing technologies, such as next-generation sequencing, enables analysis of variations between individuals at the genomics level.

Transcriptomics is the study of transcriptome that comprises the entire collection of RNA (ribonucleic acid) sequences, called transcripts, in a cell. It is estimated that a human cell contains about 25,000 transcripts. RNAs are classified into two groups: (1) mRNA is the coding RNA that is translated into protein sequences. (2) Non-coding RNAs are also classified into two subgroups; short non-coding RNAs such as microRNA (miRNA) and long non-coding RNAs (lncRNA). Non-coding RNAs are involved in gene regulation. Next-generation RNA sequencing technologies allow deeply understanding of variations and gene expression on various types of RNA molecules including miRNA, mRNA, and lncRNA [2].

Proteomics is the study of proteome, which is defined as the set of all expressed proteins and interacting protein family networks, and biochemical pathways in a cell, tissue, or organism. Although, the exact number of proteins/peptides is still unclear, it is estimated to be around a few hundred thousand.

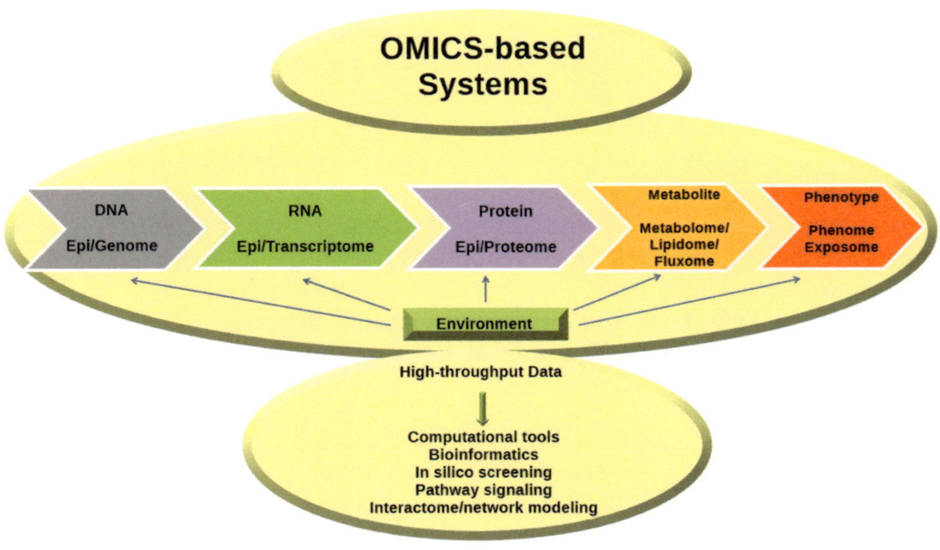

Figure 1.
Building blocks of OMICS approach and systems biology in molecular medicine.

Metabolomics is the study of metabolome within cells, biofluids, tissues, or organisms. Metabolome can be defined as the small molecules and their interactions within a biological system under a given genetic, nutritional, and environmental condition. Since the metabolome is the final downstream product, changes and interactions between gene expression, protein expression, and the environment are directly reflected in metabolome making it more physically and chemically complex than the other "omes." The metabolome is the closest to the phenotype among other omics approaches. Metabolomics best modulates and represents the molecular phenotype of health and disease [7]. In this regard, metabolomics is a brilliant source for disease-associated biomarkers. Mass spectrometry-based metabolomics/lipidomics provides a useful approach for both identification of disease-related metabolites in biofluids or tissue, and also encompasses classification and/or characterization of disease- or treatment-associated molecular patterns generated from metabolites [8, 9]. Metabolomics analysis identifies different metabotypes of disease severity and makes successful clinical and molecular phenotyping and patient stratification.

3. Application field of OMICS technology in molecular medicine

Omics-based approaches have been significantly improved recently with the addition of novel concepts such as exposome/exposomics, the study of the environmental exposure, to unravel the role of the environment in human diseases. Furthermore, the addition of adductomics, the study of compounds that bind DNA and cause damage and mutations, and volatilomics, the study of volatile organic compounds to the metabolomics/lipidomics analysis for comprehensive research of the metabolome have been newly emerging [2, 10, 11]. Exposome is a person's total lifestyle and environmental exposures, which is not well understood yet. Researchers from NIH, Dr. Chao Jiang and his colleagues, have developed a method to capture and map an individual's "exposome"—under the concept "exposing the human exposome—every breath you take, exposome tells where you have been and when." Furthermore, they have designed a portable, battery-powered device comprising sensors, a collection container with filter, and a pump that simulates human breathing to be able to track and quantify personal environmental exposures. The sensors can detect different particles such as biologicals (biotics), chemicals (abiotics), tobacco smoke, and automobile fumes. They have detected more than 2500 species, including bacteria, fungi, plants, metazoa, and more than 200 viruses. One of them was remarkably called "brochosome" which look like viral particles, in a sense, but it is actually some sort of hydrophobic protein/lipid mixture made by insects as a waterproof mechanism on their body.

Systems biology, can be defined as the integration of omics-based systems, is a hypothesis-generating approach, while classical biology is hypothesis-driven [6, 12–14]. Bioinformatics is the application of computational tools and analysis used to capture, store, and interpret biological data. Focusing on large-scale data/information obtained from a comprehensive, or global, assessment of a set of molecules, bioinformatics tools are then used to analyze the multi-dimensional amount of data to reveal metabotype, proteotype, and DNA-RNA panel biosignatures.

Analysis of multi-omics-based technologies through systems biology, bioinformatics, and computational power allows us to understand diversity of diseases, molecular heterogeneity of complex pathologies, mechanism involved in disease progression, and drug resistance. Subsequently, improvement has been made in the development of molecular-based screening, early detection, and monitoring systems as well as personalized treatment strategies [15, 16]. Omics-based integrative identification and characterization of biomarker targets and their clinical translations are essential to develop comprehensive profiling, risk stratification, future cell-targeted early interventional and therapeutic strategies. First established, a decade ago, "multi-omics" approach to disease by integrative analysis of "single omics platforms" have been a paradigm shift attributed to personalized medicine [4, 15, 17, 18]. In this manner, Chakraborty and colleagues successfully documented "onco-multi-omics" approach in cancer research [17]. Systems biology integrated high-throughput multi-omics approach has been dedicated to understand complete molecular biosignature of health and disease.

Accurate determination and validation of disease-related biomarkers necessitates the development of biorepository systems with a large collection and storage of patient biospecimens such as tissue, blood, and other bodily fluids, and well annotated clinical and pathological data [19–21]. By this way, biorepository systems enable integration of basic, translational, and clinical research to lead the discovery of hindered relevant biomarkers and emerging personalized diagnostic/therapeutic strategies on reliable big sample sizes associated with specific diseases [19, 20]. In another aspect, a recent Nature editorial (2019) critically highlights focusing on to study healthy individuals biobanking rather than people with diseases to better understand the exact definition of health with all its manifestations [22]. Projects such as "100K Wellness Project" and "The All of Us Research Program" have been producing next-generation sequencing data through specimens from healthy individuals to obtain molecular, lifestyle, and environmental measurements (http://allofus.nih.gov/), in particular for future drug discovery studies.

Genomic diversity and molecular heterogeneity of complex diseases obscure the discovery of theranostic, prognostic, and predictive biomarkers as well as their translation into personalized medicine at the single-cell level. In this aspect, promising single-cell studies formed another emerging concept in the field of the molecular medicine. Single-cell level analysis has been suggested to be crucial for a better and precise enrichment of biomarkers related to complex heterogeneous nature of diseases [23]. Omics-based analysis at the single-cell level comprises epi/genomics, epi/transcriptomics, epi/proteomics, and metabolomics/lipidomics approaches. These technologies facilitated our understanding of variations, interactions, biological functions, and disease heterogeneity at the single-cell level which paves the way for a personalized medicine-based smart healthcare system [24, 25]. Lately, one of the hottest research fields emerged as molecular characterization of circulating biomarkers composed of circulating tumor cells (CTCs), cell free DNA (cfDNA) and/or exosomes as liquid biopsies to assess disease management and evolution in real time [26]. Exosomes have been described as microvesicules (50–150 nm) released into the extracellular region by a variety of cells. Exosomes contain intact oligonucleotides, protein, and metabolites and have been identified in a vast range of biofluids including serum, urine, plasma, breast milk, saliva, pleural effusions, bronchoalveolar lavage fluid, ocular samples, tears, nasal lavage fluid, semen, synovial fluid, amniotic fluid, and pregnancy-associated serum [27]. With the development of high-throughput omics technologies, liquid biopsy has settled in the center of non-invasive or minimally invasive applications of easily accessible biofluids to detect disease-associated CTCs for diagnostic, monitoring, and therapeutic approaches. Isolation, detection, and molecular characterization of CTCs have been performed in a variety of diseases mostly in cancers. Due to high heterogeneity and resistance to treatment observed in tumor biology, single-cell CTC characterization allows clinical profiling and targeted treatment strategies and monitoring.

Molecular medicine applications not only improved the basic understanding of disease mechanism, but also contributed to the understanding of mechanism of drug action, identification of theranostic targets, and hence a paradigm shift in drug discovery [28]. Molecular theranostics can be defined as integration of disease diagnosis and treatment with the same molecular target. Promising oligonucleotide-based (DNA or RNA) therapeutics and vaccines such as gene therapy, DNA vaccines, and RNA pharmaceuticals have been successfully developed in the last 2 decades using antibodies and aptamers. Regarding DNA, viral or bacterial vectors are used and polymeric materials such as poly lactic-co-glycolic acid (PLGA), chitosan, and polyethylenimine (PEI) have been applied for efficient delivery [29]. Aptamers or antibodies can be conjugated to theranostic biomarkers and nanomaterials for specific targeting [30]. Aptamer-based applications include imaging, targeted drug delivery, and treatment such as targeted phototherapy, gene therapy, and chemotherapy [31]. Limitations in non-toxic specific targeting and delivery encouraged researchers to use drug carriers such as liposomes and nanoparticles for encapsulation of oligonucleotide therapeutics [32]. Studies on some tumor types including lung, pancreas, and breast have demonstrated successful results with encapsulated antisense oligonucleotides [33, 34]. RNA oligonucleotides using the antisense gene silencing technology has given promising results to inhibit disease-related mRNA gene expression. RNA therapeutics including antisense RNA, small interfering RNA (siRNA), and anti-miRNA (anti-miR) are promising for the treatment of a number of diseases including chronic complex diseases. Furthermore, their impact has been evaluated in the different stages of development from preclinical to Phase III clinical trials [35–39]. Major challenges dealing with efficient delivery include biocompatibility, protection from nucleases, distribution location, and persistence. Peter and colleagues have identified suicide/killer RNA molecules (siRNA, shRNA, miRNA, siRNA+miRNA complex) on numerous cancer types. In addition, they have shown that specific toxic RNAi-active sequences present in the genome can kill cancer cells [40–44]. Rozowsky and colleagues have generated a comprehensive analytic platform for extracellular RNA profiling called "exceRpt" [45].

Murillo and colleagues have created exRNA Atlas Analysis, and explored how RNA transmits information through cell-to-cell communication, known as extracellular RNA or exRNA [46]. Moreover, they have identified complexity in steps of transport exRNA molecules, types, carriers between cells, target cells, and functions, and found that even the type of carrier affected how exRNA messages were sent and received which may suggest potential novel disease-associated biomarkers and therapeutic targets. To date, exRNA-originated potential biomarkers have been identified in 13 biofluids like plasma, saliva, and urine in over 50,000 samples from over 2000 donors for nearly 30 diseases including cardiovascular diseases, diseases of the brain and central nervous system, pregnancy complications, glaucoma, diabetes, autoimmune diseases, and multiple types of cancer. Thus, exRNA profiles could be an individualized source and for personalized treatment of various diseases.

Examples of current and future applications in molecular medicine may also include DNA/RNA chips, peptide/antibody arrays, aptamer/antikor-based immunoassays, and/or sensor systems for disease screening, diagnosis, and monitoring. Molecular tools/devices such as lab-on-chips combined with sensors using microarray techniques have been developed which are able to perform patient stratification based on specified clinical and molecular features [47]. Those tools are assessed to capture very low concentrations of biochemical substances at the early disease phase, and result in effective/sensitive treatment and eradicate and/or reduce over-/undertreatment, and side effects [48–50].

4. Conclusions

In contrast to one single gene disease concept of Mendelian inheritance, chronic complex diseases are result of alterations in multiple genes and signaling pathways. Furthermore, these diseases are generally characterized with heterogeneity at the cellular/tissue level. Therefore, identification and omics-based profiling of multiple biomarker profiles rather than one single gene/biomarker possess greater statistical power and reliability for future screening/diagnosis/monitoring/treatment strategies. In this aspect, molecular medicine applications have brought novel and significant outputs to the research as well as challenges that require further preclinical and clinical studies. Development of omics-based discriminatory biomarkers for early detection, as well as novel targeted interventional and therapeutic strategies are crucial for a personalized healthy life as well as disease management.

Author details

Sinem Nalbantoglu* and Abdullah Karadag*
TUBITAK Marmara Research Center, Gene Engineering and Biotechnology Institute, Molecular Oncology Laboratory, Gebze, Kocaeli, Turkey

*Address all correspondence to: nalbantoglusinem@gmail.com and abdullah.karadag@tubitak.gov.tr

References

[1] Crick F. Central dogma of molecular biology. Nature. 1970;**227**:561-563. DOI: 10.1038/227561a0

[2] Holland N. Future of environmental research in the age of epigenomics and exposomics. Reviews on Environmental Health. 2017;**32**(1-2):45-54. DOI: 10.1515/reveh-2016-0032

[3] Yadav SP. The wholeness in suffix -omics, -omes, and the word om. Journal of Biomolecular Techniques. 2007;**18**(5):277

[4] Hasin Y, Seldin M, Lusis A. Multi-omics approaches to disease. Genome Biology. 2017;**18**(1):83. DOI: 10.1186/s13059-017-1215-1

[5] Aardema MJ, MacGregor JT. Toxicology and genetic toxicology in the new era of "toxicogenomics": Impact of "-omics" technologies. Mutation Research. 2002;**499**(1):13-25. DOI: 10.1016/S0027-5107(01)00292-5

[6] Horgan RP, Kenny LC. 'Omic' technologies: Genomics, transcriptomics, proteomics and metabolomics. The Obstetrician & Gynaecologist. 2011;**13**:189-195. DOI: 10.1576/toag.13.3.189.27672

[7] Guijas C, Montenegro-Burke JR, Warth B, Spilker ME, Siuzdak G. Metabolomics activity screening for identifying metabolites that modulate phenotype. Nature Biotechnology. 2018;**36**(4):316-320. DOI: 10.1038/nbt.4101

[8] Gowda GA, Zhang S, Gu H, Asiago V, Shanaiah N, Raftery D. Metabolomics-based methods for early disease diagnostics. Expert Review of Molecular Diagnostics. 2008;**8**(5):617-633. DOI: 10.1586/14737159.8.5.617

[9] Vander Heiden MG. Targeting cancer metabolism: A therapeutic window opens. Nature Reviews. Drug Discovery. 2011;**10**(9):671-684. DOI: 10.1038/nrd3504

[10] Broza YY, Zuri L, Haick H. Combined volatilomics for monitoring of human body chemistry. Scientific Reports. 2014;**4**:4611. DOI: 10.1038/srep04611

[11] Taware R, Taunk K, Pereira JAM, Shirolkar A, Soneji D, Câmara JS, et al. Volatilomic insight of head and neck cancer via the effects observed on saliva metabolites. Scientific Reports. 2018;**8**(1):17725. DOI: 10.1038/s41598-018-35854-x

[12] Kell DB, Oliver SG. Here is the evidence, now what is the hypothesis? The complementary roles of inductive and hypothesis-driven science in the post-genomic era. BioEssays. 2004;**26**:99-105. DOI: 10.1002/bies.10385

[13] Kell DB. The virtual human: Towards global systems biology of multiscale, distributed biochemical network models. Life. 2007;**59**:689-695. DOI: 10.1080/15216540701694252

[14] Westerhoff HV, Palsson BO. The evolution of molecular biology into systems biology. Nature Biotechnology. 2004;**22**:1249-1252. DOI: 10.1038/nbt1020

[15] Santolini M, Romay MC, Yukhtman CL, Rau CD, Ren S, Saucerman JJ, et al. A personalized, multiomics approach identifies genes involved in cardiac hypertrophy and heart failure. npj Systems Biology and Applications. 2018;**4**:12. DOI: 10.1038/s41540-018-0046-3

[16] Trock BJ. Application of metabolomics to prostate cancer. Urologic Oncology. 2011;**29**(5):572-581. DOI: 10.1016/j.urolonc.2011.08.002

[17] Chakraborty S, Hosen MI, Ahmed M, Shekhar HU. Onco-multi-omics approach: A new frontier in cancer research. BioMed Research International. 2018;**2018**:9836256. DOI: 10.1155/2018/9836256

[18] Hu Y, An Q, Sheu K, Trejo B, Fan S, Guo Y. Single cell multi-omics technology: Methodology and application. Frontiers in Cell and Development Biology. 2018;**6**:28. DOI: 10.3389/fcell.2018.00028

[19] Hewitt RE. Biobanking: The foundation of personalized medicine. Current Opinion in Oncology. 2011;**23**(1):112-119. DOI: 10.1097/CCO.0b013e32834161b8

[20] Lee JE, Kim YY. Impact of preanalytical variations in blood-derived biospecimens on omics studies: Toward precision biobanking? OMICS. 2017;**21**(9):499-508. DOI: 10.1089/omi.2017.0109

[21] Liu A, Pollard K. Biobanking for personalized medicine. Advances in Experimental Medicine and Biology. 2015;**864**:55-68. DOI: 10.1007/978-3-319-20579-3_5

[22] Editorial. Banking on health. Nature Biotechnology. 2019;**37**(3):197. DOI: 10.1038/s41587-019-0069-3

[23] Niu F, Wang DC, Lu J, Wu W, Wang X. Potentials of single-cell biology in identification and validation of disease biomarkers. Journal of Cellular and Molecular Medicine. 2016;**20**(9): 1789-1795. DOI: 10.1111/jcmm.12868

[24] Mannello F, Ligi D, Magnani M. Deciphering the single-cell-Omic: Innovative application for translational medicine. Expert Review of Proteomics. 2012;**6**:635-648. DOI: 10.1586/epr.12.61

[25] Mannello F. Single-cell analysis: From innovative omics to target therapy. Journal of Pharmacogenomics and Pharmacoproteomics. 2012;**3**:6. DOI: 10.4172/2153-0645.1000e130

[26] Jiang Y, Wang D. Liquid biopsy in the OMICS era of tumor medicine. Open Access Journal of Biomedical Engineering and Its Applications. 2018;**1**(3):115. DOI: 10.32474/OAJBEB.2018.01

[27] Halvaei S, Daryani S, Eslami-S Z, Samadi T, Jafarbeik-Iravani N, Bakhshayesh TO, et al. Exosomes in cancer liquid biopsy: A focus on breast cancer. Molecular Therapy--Nucleic Acids. 2018;**10**:131-141. DOI: 10.1016/j.omtn.2017.11.014

[28] Sansonetti PJ. Moving molecular medicine. EMBO Molecular Medicine. 2017;**9**(4):395. DOI: 10.15252/emmm.201707746

[29] Lee DY, Li KC. Molecular theranostics: A primer for the imaging professional. AJR. American Journal of Roentgenology. 2011;**197**(2):318-324. DOI: 10.2214/AJR.11.6797

[30] Xiang D, Zheng C, Zhou SF, Qiao S, Tran PH, Pu C, et al. Superior performance of aptamer in tumor penetration over antibody: Implication of aptamer-based theranostics in solid tumors. Theranostics. 2015;**5**:1083-1097. DOI: 10.7150/thno.11711

[31] Pang X, Cui C, Wan S, Jiang Y, Zhang L, Xia L, et al. Bioapplications of cell-SELEX generated aptamers in cancer diagnostics, therapeutics, theranostics and biomarker discovery: A comprehensive review. Cancers (Basel). 2018;**10**(2):47. DOI: 10.3390/cancers10020047

[32] Urban-Klein B, Werth S, Abuharbeid S, Czubayko F, Aigner A. RNAi-mediated gene-targeting through systemic application of polyethylenimine (PEI)-complexed siRNA in vivo. Gene Therapy. 2005;**12**(5):461-466. DOI: 10.1038/sj.gt.3302425

[33] Pan B, Cui D, Sheng Y, Ozkan C, Gao F, He R, et al. Dendrimer-modified magnetic nanoparticles enhance efficiency of gene delivery system. Cancer Research. 2007;**67**(17): 8156-8163. DOI: 10.1158/0008-5472. CAN-06-4762

[34] Dobson J. Magnetic micro- and nano-particle-based targeting for drug and gene delivery. Nanomedicine (London, England). 2006;**1**(1):31-37. DOI: 10.2217/17435889.1.1.31

[35] Huang YF, Shangguan D, Liu H, Phillips JA, Zhang X, Chen Y, et al. Molecular assembly of an aptamer-drug conjugate for targeted drug delivery to tumor cells. ChemBioChem. 2009;**10**: 862-868. DOI: 10.1002/cbic.200800805

[36] Cabarcas S, Watabe K, Schramm L. Inhibition of U6 snRNA transcription by PTEN. OnLine Journal of Biological Sciences. 2010;**10**(3):114-125. DOI: 10.3844/ojbsci.2010.114.125

[37] Schiffelers RM, Ansari A, Xu J, Zhou Q, Tang Q, Storm G, et al. Cancer siRNA therapy by tumor selective delivery with ligand-targeted sterically stabilized nanoparticle. Nucleic Acids Research. 2004;**32**(19):e149. DOI: 10.1093/nar/gnh140

[38] Harper SQ. Progress and challenges in RNA interference therapy for Huntington disease. Archives of Neurology. 2009;**66**(8):933-938. DOI: 10.1001/archneurol.2009.180

[39] Pai SI, Lin YY, Macaes B, Meneshian A, Hung CF, Wu TC. Prospects of RNA interference therapy for cancer. Gene Therapy. 2006;**13**(6):464-477. DOI: 10.1038/sj.gt.3302694

[40] Murmann AE, McMahon KM, Haluck-Kangas A, Ravindran N, Patel M, Law CY, et al. Induction of DISE in ovarian cancer cells in vivo. Oncotarget. 2017;**8**(49):84643-84658. DOI: 10.18632/ oncotarget.21471

[41] Patel M, Peter ME. Identification of DISE-inducing shRNAs by monitoring cellular responses. Cell Cycle. 2018;**17**(4):506-514. DOI: 10.1080/15384101.2017.1383576

[42] Putzbach W, Gao QQ, Patel M, Haluck-Kangas A, Murmann AE, Peter ME. DISE: A seed-dependent RNAi off-target effect that kills cancer cells. Trends in Cancer. 2018;**4**(1):10-19. DOI: 10.1016/j.trecan.2017.11.007

[43] Putzbach W, Gao QQ, Patel M, van Dongen S, Haluck-Kangas A, Sarshad AA, et al. Many si/shRNAs can kill cancer cells by targeting multiple survival genes through an off-target mechanism. eLife. 2017;**6**:e29702. DOI: 10.7554/eLife.29702

[44] Putzbach W, Haluck-Kangas A, Gao QQ, Sarshad AA, Bartom ET, Stults A, et al. CD95/Fas ligand mRNA is toxic to cells. eLife. 2018;7. DOI: 10.7554/ eLife.38621

[45] Rozowsky J, Kitchen RR, Park JJ, Galeev TR, Diao J, Warrell J, et al. exceRpt: A comprehensive analytic platform for extracellular RNA profiling. Cell Systems. 2019;**S2405-4712**(19):30074-30072. DOI: 10.1016/j. cels.2019.03.004

[46] Murillo OD, Thistlethwaite W, Rozowsky J, Subramanian SL, Lucero R, Shah N. exRNA atlas analysis reveals distinct extracellular RNA cargo types and their carriers present across human biofluids. Cell. 2019;**177**(2):463.e15-477. e15. DOI: 10.1016/j.cell.2019.02.018

[47] Zhang Q, Zhang M, Djeghlaf L, Bataille J, Gamby J, Haghiri-Gosnet AM, et al. Logic digital fluidic in miniaturized functional devices: Perspective to the next generation of microfluidic lab-on-chips. Electrophoresis. 2017;**38**(7):953-976. DOI: 10.1002/elps.201600429

[48] Donner A. Nanotechnology in molecular medicine. Trends in

Molecular Medicine. 2010;**16**(12): 551-552. DOI: 10.1016/j.molmed.2010.10.001

[49] Boenink M. Molecular medicine and concepts of disease: The ethical value of a conceptual analysis of emerging biomedical technologies. Medicine, Health Care, and Philosophy. 2010;**13**(1):11-23. DOI: 10.1007/s11019-009-9223-x

[50] Wickline SA, Lanza GM. Molecular imaging, targeted therapeutics, and nanoscience. Journal of Cellular Biochemistry. Supplement. 2002;**39**: 90-97. DOI: 10.1002/jcb.10422

Chapter 2

Precision Medicine: Role of Biomarkers in Early Prediction and Diagnosis of Alzheimer's Disease

Liming Shen, Sijian Xia, Huajie Zhang, Fang Yao, Xukun Liu, Yuxi Zhao, Ming Ying, Javed Iqbal and Qiong Liu

Abstract

Alzheimer's disease (AD), the most common form of dementia in the aged people, is a chronic and irreversible neurodegenerative disorder. Early prediction, intervention, and objective diagnosis are very critical in AD. In this chapter, we will introduce the current progress in the prediction and diagnosis of AD, including recent development in diagnostic criteria, genetic testing, neuroimaging techniques, and neurochemical assays. Focus will be on some new applied methods with more specific examples, that is, cerebrospinal fluid (CSF) and blood proteins and peptides, which might serve as biomarkers for the diagnosis of AD. We will also discuss biomarker-based diagnostic strategies and their practical application.

Keywords: Alzheimer's disease, mild cognitive impairment, diagnosis, prediction, biomarker

1. Introduction

Alzheimer's disease (AD), the most common type of dementia in aged people, is an untreatable neurodegenerative disorder characterized by abnormal accumulations of amyloid-β (Aβ) oligomers and intracellular neurofibrillary tangles (NFTs) in the brain attributable to hyperphosphorylated tau that results in progressive synaptic dysfunction and cognitive deficits [1, 2]. AD is the fifth leading cause of death for people aged 65 and over [3] and is officially listed as the sixth leading cause of death in the United States [4]. Presently, more than 47 million people are estimated to be living with dementia globally, and this number is projected to rapidly increase, reaching 75 million by 2030 and 135 million by 2050 [5]. In China, there were 135.2 million aged people in 2015, and 8.5 million of them were oldest-old (beyond 85 years and above). Based on age-specific prevalence of AD, China would have over 20 million AD patients in 2050 [6].

AD is a heterogeneous disease caused by a combination of environmental and genetic factors. Currently known risk factors for AD include age, sex, cardiovascular risk factors and metabolic risk factors, sleep apnea, family history, and certain genetic variants [7]. In an attempt to explain the complexity and multifactorial nature of AD, various hypotheses are established. These include Aβ aggregation, tau aggregation, metal dyshomeostasis, oxidative stress, cholinergic dysfunction, inflammation, and downregulation of autophagy [8]. However, none of the hypotheses is capable of independently explaining the pathological conditions observed in AD. The amyloid cascade hypothesis is widely considered to be involved in the pathogenesis of AD [9]. The anatomic and temporal discordance between Aβ pathology, tau aggregation, and neurodegeneration has led to the postulation of Aβ being an initiator of a complex cascade that ends in tau-medicated neurodegeneration [7].

As the etiology and pathogenesis of AD have not been elucidated, none of the proposed pharmacologic treatments (medications) are authentic to slow or stop the neurodegeneration [10]. On the other hand, in clinical practice, a diagnosis of AD is primarily made on the base of clinical features, results of neurological and neuropsychological tests, and by exclusion of other causes of dementia, including vascular and frontotemporal dementia or other neurological diseases [11]. Although a variety of imaging techniques, and detection of levels of $A\beta_{40}$, $A\beta_{42}$, total tau protein (T-tau), and phosphorylated tau protein (P-tau) in CSF have been found to be able to support clinical diagnosis of mild cognitive impairment (MCI) and AD. However, CSF collection is invasive, and therefore, its sampling is quite difficult. Imaging techniques are expensive, which restrict their application either as routine screening tools or for repetition of tests to monitor the drug treatment or pathological progress [12].

The goal of precision medicine is to use biological knowledge and other related health information to predict individual disease risk, understand disease etiology, identify disease subcategories, improve diagnosis, and provide personalized treatment strategies [7]. To date, none of the effective intervention is available, which can cure or halt the progression of AD. However, studies have consistently shown that active management of Alzheimer's and other dementias can improve the quality of life of affected subjects and their caregivers [4]. The development of biomarkers for AD is making it possible to detect the disease and provide an accurate diagnosis earlier, which is beneficial for diagnosed individuals, their caregivers and loved ones, as well as society as a whole [13]. In particular, pathophysiological alterations associated with AD are thought to begin several decades before the onset of the disease [14]. Thus, early diagnosis would provide a crucial opportunity for intervention in AD progression. In addition, the use of biomarkers in all stages of Alzheimer's disease will facilitate to develop therapeutic strategy that targets the underlying brain changes at each stage. Moreover, the research of biomarkers discovery may contribute to enhance our understanding of the pathogenesis of the disease itself.

Therefore, the biomarker discovery is of utmost importance to improve diagnostics and prevention of disease and to monitor treatment effects. In this chapter, we introduced the update of AD diagnostic criteria, genetic research, and imaging and fluid (CSF and blood) biomarkers, highlighting the progress of biomarker research and advances in methodology.

2. Advance in diagnostic criteria

2.1 Development of AD biomarkers in early guidance

The first set of criteria proposed for diagnosis of AD was launched in 1984 by a workgroup from the National Institute of Neurological and Communicative Disorders and Stroke and the Alzheimer's disease and Related Disorders Association

(NINCDS–ADRDA) [15]. At that time, Alzheimer's pathological changes could not be measured in vivo. The criteria focused on clinical symptoms and required the presence of significant disability and impact on daily living. Thus, it allowed only a "probable" diagnosis of AD to be reached while the person was alive, and a definitive diagnosis could be made only if Alzheimer's pathology was found at autopsy [15]. In terms of distinguishing AD from other types of dementia, its specificity was low [16]. In 2007, the AD Research Diagnostic Criteria proposed by the International Working Group (IWG) first included true biomarkers in the criteria for active diagnosis of AD and considered AD to progress from preclinical, pre-dementia to dementia [17].

In 2011, the National Institute on Aging and the Alzheimer's Association (NIA-AA) proposed revised guidelines for diagnosing AD [18–21]. This updated diagnostic criteria and guidelines incorporated biomarker tests in addition to clinical symptoms, and provided the researchers tools for diagnosing AD. It identified AD as a continuum with three distinct stages: preclinical stage, mild cognitive impairment (MCI), and dementia. There are two types of biomarker, the former including CSF $A\beta_{42}$ or amyloid positron emission tomography (PET) and the latter including CSF tau/P-tau, MRI hippocampus or medial axillary atrophy, and low glucose metabolism on PET or SPECT [18]. In 2012, the NIA-AA also developed new guidelines to help pathologists to describe and categorize the brain changes associated with Alzheimer's and other dementias on autopsy [22]. Parallel to the hypothetical pathophysiological sequence of AD, a biomarker model was proposed by Jack et al. [23, 24]. It revealed that the biomarker abnormality occurs first in Aβ levels, which can either be in the form of an upregulation in plasma or down-regulation in CSF in usually cognitively normal individuals and can be detected by biochemical analysis [6, 23, 24]. The next stage of biomarker abnormality was usually amyloid deposition in the brain detected by abeta PET. Subsequently, the changes of biomarkers include neuronal injury, indicated by increased levels of CSF total phosphorylated tau proteins, and cerebral atrophy revealed by structural MRI, as well as neurodegeneration and synaptic dysfunction detected by reduced fluorodeoxyglucose (FDG) uptake through PET. These results of biomarker studies showed that they are correlated with different disease stages, which thus correlate with and support the changes in the "abeta hypothesis" [6]. Furthermore, research criteria for diagnosing preclinical states of AD developed by the International Working Group (the IWG-2 criteria) require the individual to be asymptomatic and have a marker of AD pathology or an AD autosomal dominant mutation on chromosome 1, 14, or 21 [16]. This is describing an at-risk state where progression of AD is not inevitable.

2.2 NIA-AA 2018 update guidelines

Recently, NIA-AA guidelines for AD have been updated [25]. In these latest guidelines, Alzheimer's biomarkers are divided into three categories (the A/T/N system). The classification uses three types of biomarkers as shown in **Table 1**. "A" refers to amyloid β (Aβ) as measured either by amyloid PET imaging of amyloid plaques or in the cerebrospinal fluid (CSF) as $A\beta_{42}$ or the $A\beta_{42}$ to $A\beta_{40}$ ratio. "T" refers to tau pathology as measured by CSF phosphorylated tau or tau PET imaging of parenchymal neurofibrillary tangles. "N" refers to neurodegeneration or neuronal injury and dysfunction, as measured by elevated levels of CSF total tau, decreased glucose metabolism shown on FDG-PET imaging, and brain atrophy shown with structural MRI. While "A" and "T" are considered to have diagnostic specificity for AD, "N" is not specific for AD diagnoses because it can reflect any number of etiologies in addition to AD. The A/T/N biomarkers may reflect the

Biomarker class	CSF marker	Imaging marker
Amyloid (A)	CSF $A\beta_{42}$,or $A\beta_{42}/A\beta_{40}$ ratio	Amyloid PET
Tau (T)	CSF phospho-tau	Tau PET
Neurodegeneration (N)	CSF total tau	Anatomic MRI; FDG PET

Table 1.
AT(N) biomarker grouping of the NIA-AA Framework.

presence (state) or progression (stage) of a disease. An individual with biomarker evidence of Aβ deposition alone (abnormal amyloid PET scan or low CSF $A\beta_{42}$ or $A\beta_{42}/A\beta_{40}$ ratio) with a normal pathologic tau biomarker would be assigned the label "Alzheimer's pathologic change" [25]. The term "Alzheimer's disease" would be applied if biomarker evidence of both Aβ and pathologic tau is present. Alzheimer's pathologic change and AD are not regarded as separate entities but earlier and later phases of the "Alzheimer's continuum" (an umbrella term that includes both). These definitions are applied independently from clinical symptoms [25].

In addition, together with cognitive symptoms (C), AT (N)(C) measures have different roles for definition and staging, A and T indicate specific neuropathologic changes that define Alzheimer's disease, that is, Aβ biomarkers determine whether or not an individual is in the Alzheimer's continuum. Pathologic tau biomarkers determine if someone who is in the Alzheimer's continuum has Alzheimer's disease. (N) and (C) are not specific to AD and are therefore placed in parentheses. They indicate staging severity [25].

NIA-AA 2018 guidelines are still research framework and cannot be considered as routine clinical care [25]. However, clearly, with the update of these guidelines, the definition of AD shifts from symptom-based definition to biology-based definition. This is leading to a better understanding of the underlying mechanisms of the disease and aiding in the development of new interventions to delay or prevent disease progression and biomarker research [25].

3. Genetic susceptibility

AD can be divided into early-onset familial AD (EOAD) and late-onset AD (LOAD). Early-onset AD accounts for less than 1–5% and is caused by highly penetrable variants, the majority of which are attributable to mutations in one of the three genes, amyloid precursor protein (APP), presenilin 1 (PSEN1), and presenilin 2 (PSEN2) [26, 27]. Individuals with Alzheimer's mutations in any of these three genes tend to develop symptoms before age 65, and the average age is 50 years [28]. More than 95% of AD cases are LOAD, which are "sporadic" with no apparent familial recurrence of the disease and are caused by a more complex underlying genetic architecture, and typically appear in older individuals (age 65 years and over) [29].

E4 allele of apolipoprotein E (APOE4) is the only verified genetic risk factor for late-onset AD. It is present in approximately 15% of the normal population; however, it occurs in 50% of those developing AD. APOE encodes a lipid carrier apolipoprotein E (ApoE) that is found both in the periphery and the central nervous system [7]. APOE4 shows the complex interplay of mechanisms contributing to sporadic AD, including reduced cholesterol transport, less efficient Aβ clearance and more aggregation, triggering neurotoxicity through tau phosphorylation, increased brain neuronal activity and atrophy, reduced synaptic plasticity, and greater neuroinflammation [7].

In addition to APOE4, genome-wide association study (GWAS) has identified more than 30 genomic loci that are associated with AD risk [7]. Ten susceptible loci for LOAD with the most consistent results include APOE, CLU, PICALM, CR1, BIN1, EPHA1, MS4A, ABCA7, CD33, and CD2AP. These AD risk loci are associated with different biological processes, including immune system, endocytosis, lipid homeostasis, and Aβ metabolism, highlighting the complexity of AD and point toward potentially novel directions for drug discovery and treatment [30]. Besides, rare variants (allele frequency < 1%) that influence the risk for LOAD have also been identified in several genes, including TREM2, PLD3, UNC5C, AKAP9, ADAM10, and ABI3 [7].

Genetics can provide a valuable starting point for advancement. To date, the vast majority of genetic work in AD has been the search for individual genes or combinations of genes associated with a dichotomous outcome of an AD diagnosis. For example, a study used survival analysis modeling to integrate AD risk variants and develop a polygenic hazard score for age of onset, which show a strong genetic component in AD that can be useful in predicting risk. Thus, genetic knowledge may also facilitate precision medicine. This approach has recently been proposed for dementia [7].

4. Biomarkers of AD

4.1 Imaging

As mentioned above, three Alzheimer's neuroimaging biomarkers are currently used for research and, in some cases, are used to aid in clinical diagnosis. Elevated cortical tau shown with PET imaging is a biomarker for neurofibrillary tangles; decreased glucose metabolism shown by FDG-PET imaging and atrophy shown by structural MRI are biomarkers for neurodegeneration or neuronal injury [31]. Deposition of Aβ can be detected by amyloid-specific imaging agents for positron emission tomography-computed tomography (PET/CT) as early as 15 years before the onset of AD symptoms, whereas the next most sensitive metric, cerebral hypometabolism (FDG-PET/CT) is detectable only 10 years prior to symptom onset. Aβ PET/CT is thought to precede by 10 years the declines in even the most sensitive cognitive metrics including episodic memory [32].

4.2 Fluid biomarkers for AD

In parallel to imaging biomarkers, additional types of biomarkers currently being studied in AD and used mainly for research purposes are found in CSF and blood.

4.2.1 CSF biomarker

The most validated CSF biomarkers for AD are $A\beta_{42}$, total tau (T-tau), and tau phosphorylated at threonine 181 (P-tau181) [33]. These biomarkers have consistently shown a marked change in AD dementia and also in the early prodromal phase of the disease. In CSF of AD patients, a decreased level of $A\beta_{42}$ has consistently been found, whereas the concentrations of tau and P-tau are increased [33]. The levels of CSF tau and P-tau have been found to correlate with brain atrophy in AD, while a reduction of $A\beta_{42}$ in CSF is shown to correlate with brain atrophy in non-demented subjects indicating a potential preclinical stage [34]. In addition, high CSF T-tau and P-tau predict the progression of cognitive symptoms better than $A\beta_{42}$ during a clinically relevant time period (1–2 years) [35]. Based on their high diagnostic performance, as state above, these core AD CSF biomarkers have been included in the diagnostic criteria for AD [18, 25]. However, CSF biomarkers show 20–30% interlaboratory and interassay variability [36]. In order to reduce this variability, standardization efforts include the creation of a mass spectrometry (MS)-based reference measurement procedures (RMP) for CSF $A\beta_{42}$ [37] and certified reference materials (CRM) for the main AD CSF biomarkers [38]. Precise measurements have also been achieved by novel assays developed on fully automated laboratory equipment [39]. Moreover, other Aβ protein levels and ratios (tau/$A\beta_{42}$, $A\beta_{42}/A\beta_{40}$, $A\beta_{42}/A\beta_{38}$) also become abnormal with the signature of AD [40]. For example, reduced $A\beta_{42}/A\beta_{40}$ ratio is characteristic of AD dementia and prodromal AD [41].

Despite the promising CSF core biomarkers for the identification of presymptomatic AD and discriminate AD cases well from healthy subjects, the inherent heterogeneity in the progression of mounting plaque and tangle load over time between patients, as well as the presence of mixed pathologies and different comorbidities, are considered [42]. For example, elevated amyloid deposition is frequently found in cognitively normal subjects, and CSF levels of Aβ and Aβ imaging with PIB-PET do not correlate with cognitive decline [43]. Thus, it is needed to augment the CSF core biomarkers with novel proteins to improve diagnostic accuracy in longitudinal studies [44]. Recently, new biomarkers reflecting other aspects of pathophysiology have been reported, for example, CSF neurofilament light chain (NFL), neurogranin, and YKL-40 proteins have reached at an advanced clinical validation stage [45]. A recent meta-analysis showed that the core CSF biomarkers of neurodegeneration (T-tau, P-tau, and $A\beta_{42}$) and CSF NFL were strongly associated with AD, and NSE, VLP-1, HFABP, and YKL-40 were moderately associated with AD [33]. Among these, NFL, NSE, VLP-1, and HFABP are related to neurodegeneration, and YKL-40 is associated with glial activation [33]. Of note, another protein, neurogranin, involved in synaptic dysfunction and degeneration, is found with higher CSF levels in patients with AD. It is seemingly specific for AD and does not change in the majority of other neurodegenerative disorders [35]. Taken together, the integration of complementary pathophysiological biomarker candidates covering additional key AD mechanisms will likely result in an incremental performance optimization for the detection, diagnosis, and differential diagnosis of primary neurodegenerative diseases and dementia disorders [45].

4.2.2 Blood-based biomarker

Blood collection is routinely performed, minimally invasive and cheap and suitable for recurrent measures. Blood-based biomarkers may allow for efficient monitoring of disease processes in AD and could be used as a screening tool in primary care [46]. Amyloid β (Aβ) is a widely researched plasma biomarker for AD. Evidence supporting the transport of Aβ across the blood-brain barrier and through CSF suggests that 30–50% of plasma Aβ originates from the CNS [47]. However, diagnostic relevance of plasma Aβ for AD process yields conflicting results [33, 48, 49]. In terms of this, a meta-analysis found that lower $A\beta_{42}$:$A\beta_{40}$ ratios are significantly associated with the development of AD and dementia [49]. Another meta-analysis showed that plasma or serum concentration of $A\beta_{40}$ did not differ significantly between patients with AD subjects and controls [33]. Recently, using the INNO-BIA kit based on a multiplex xMAP technique, Hanon et al. found that plasma $A\beta_{42}$ and $A\beta_{40}$ are lower in AD than in amnestic MCI and non-amnestic MCI, respectively. Plasma $A\beta_{42}$ correlated with age,

Mini–Mental State Examination, and APOE ε4 allele [48]. Another AD pathology, that is, tau, a meta-analysis suggested that plasma T-tau are strongly associated with AD [33].

Indeed, CNS-specific proteins with very low concentrations in the blood are difficult to quantify using standard immunochemical technologies, such as ELISA (enzyme-linked immunosorbent assay), which is a major challenge in developing blood biomarkers [20]. This might be one of the reasons for the inconsistency of the analysis results in the previous studies. Recent technical breakthroughs in the field of ultrasensitive assays have started to improve it [50]. These technologies include single-molecule array (Simoa) technology and immunomagnetic reduction (IMR) [50]. Simoa technology can detect single protein molecules in blood, which captured target proteins on microscopic beads decorated with specific antibodies and then labeled the immunocomplexes (one- or zero-labeled target protein molecules per bead) with an enzymatic reporter capable of generating a fluorescent product. After isolating the beads in 50-fl reaction chambers designed to hold only a single bead, fluorescence imaging is detected [51]. The average sensitivity improvement of the Simoa immunoassays versus conventional ELISA was >1200-fold, with coefficients of variation of <10% [52]. By using this technique, Mattsson et al. found associations between elevated plasma tau and AD hallmarks, but these were mild and differed between cohorts, and high plasma tau is associated with rapid progression in later disease stages [53]. More recently, by using this platform, Tatebe et al. reported the quantitative data on the plasma levels of P-tau181 in controls and patients with AD and Down syndrome (DS). These data suggest that the plasma P-tau181 is a promising blood biomarker for brain AD pathology [54]. Mielke et al. reported that plasma total tau and P-tau181 levels are higher in AD dementia patients than those in cognitively unimpaired and total tau and P-tau181 levels are higher in AD dementia patients than those in cognitively unimpaired, and plasma P-tau181 are more strongly associated with both Aβ and tau PET [55]. Interestingly, the neuronal injury marker NFL mentioned above was also found to be increased in plasma of the patients with MCI and patients with AD dementia with Aβ pathologic features by using Simoa technology [46].

Another ultrahigh-sensitive technology is referred to as a superconducting quantum interference device (SQUID) immunomagnetic reduction (IMR) assay. Magnetic nanoparticles are coated with an antibody, and on binding of the analyte, the oscillation of the particles in an alternating magnetic field is decreased in a concentration-dependent manner [56]. Using the SQUID-based IMR, the low detection limit for amyloids and tau protein is found to be 1–10 pg/mL [57, 58]. Thus, it makes possible the measurement of plasma biomarkers for the diagnosis of AD [58–61]. For example, by IMR technology, the previous studies suggested that the plasma $A\beta_{42}$ is a useful biomarker for AD. The $A\beta_{42}/A\beta_{40}$ ratio improves the diagnostic power of the plasma Aβ biomarkers [58], and plasma $A\beta_{42}$ correlates with CSF $A\beta_{42}$ in AD [59]. Additional researches indicated that plasma $A\beta_{42}$ and tau can be used to assist in the clinical diagnosis of AD [60], and the concentration of P-tau181 in plasma can be used to differentiate memory disorder/cognitive decline in early-stage AD patients [61]. Clearly, these ultrahigh-sensitive assay technologies provide novel methods to measure low-level proteins especially in blood. These AD-specific proteins such as Aβ, and tau-related proteins or the protein biomarkers at low concentrations in the bloodstream for AD and may serve as clinical tools for the diagnosis of AD.

Besides, the studies of blood-based biomarkers also cover the following aspects: searching for other disease pathology related to proteins in blood; blood-based biomarker panels; and markers of inflammation, oxidative stress, mitochondrial dysfunction, and neuronal and microvascular injury [62].

5. Mass spectrometry (MS)-based methods and approaches

Numerous reports have demonstrated that MS-based methods can be robust, and accurate MS has been playing an important role in studying peptide and protein identities, structures, modifications, and interactions that collectively drive their biological functions. MS-based technology has been used to study the pathogenesis of AD and biomarkers in body fluids, such CSF, plasma, urine, and saliva.

5.1 Proteomics

MS-based proteomics technology is well suited for the biomarker discovery for diseases such as AD [63–65]. During the last 10 years, apart from the gel-based techniques (e.g., 2D-PAGE and 2D-DIGE), gel-free techniques (e.g., stable isotope labeling or using label-free methods) have been dominating the field of MS-based quantitation in proteomics [66]. Including our previous study [67], the method of iTRAQ with multidimensional liquid chromatography and tandem mass spectrometry has been used to reveal many candidate proteins as potential biomarkers of MCI or AD [67–69]. One of our quantitative proteomics-based studies revealed the differentially expressed proteins in AD subjects [67]. These proteins were found involved in various biological processes and pathways, such as Aβ metabolism, inflammatory and immune response, and oxidative stress, which have previously been reported to be linked with AD, supporting the existing theories of AD pathophysiology. Furthermore, some new technologies such as SWATH-MS will also be applied to further enhance probability of AD biomarkers. SWATH-MS is a specific further variant of data-independent acquisition (DIA) methods and is emerging as a technology that combines deep proteome coverage capabilities with quantitative consistency and accuracy [70].

Apart from quantitative proteomics, the development of assays to quantify particular post-translational modification of proteins is also being considered [65, 71, 72]. For example, the carbonylation of proteins associated with oxidative stress has been studied in AD [72]. Using Western blotting with two-dimensional gel electrophoresis (2D-Oxyblot), we investigated the specifically carbonylated proteins in the hippocampi [73] and serum [74] of triple transgenic mouse model of AD (3 × Tg-AD) at the early age of month, some carbonylated proteins were identified as significantly oxidized proteins compared with the control in both of the samples. This suggests that oxidative stress is an early event in AD progression, and these oxidized proteins in the serums may provide potential biomarkers of AD at the early stage. This is similar to two previous studies [75, 76]; where the authors observed serum protein carbonylation in MCI and found increased levels of carbonylation at this stage of cognitive decline.

Together, the proteomic approach is comparatively new and more advanced for biomarker analysis of proteins and provides a complementary way to obtain such a comprehensive data.

5.2 Targeted proteomic approaches

There are generally three different stages in the development of new biomarkers: the discovery phase (i.e., screening), the verification phase, and the validation phase. Multiple reaction monitoring (MRM), also known as selected reaction monitoring, is a targeted mass spectrometry approach to protein quantitation and is emerging to bridge the gap between biomarker discovery and clinical validation [77, 78]. Highly multiplexed MRM assays are readily configured and enable simultaneous verification of large numbers of candidates facilitating the development of biomarker panels which can increase specificity [77, 78]. MRM can enhance the lower detection limit for peptides due to its ability to rapidly and continuously monitor exclusively for the specific ions of interest. MRM analysis combine with stable isotope also offers multiplexing capability and increases the reliability of quantification [77, 78]. As AD is a multifactorial disease, a panel of proteins is more suitable as biomarker for AD. Thus, MRM is a valuable tool to verify biomarker candidates for AD and possible future practical applications. Several studies have emerged using MRM to identify CSF-based protein biomarkers of AD [79–81]. In addition to MRM, parallel reaction monitoring (PRM) technique has also been used to evaluate biomarker candidates for AD [82, 83]. PRM is related to the SRM approach but has the advantage of acquiring full fragment spectra instead of a choice of preselected fragments; interfering signals are avoided, whereas quantitation and high sensitivity are conserved [64]. In this way, other biochemical pathways and proteins which are not directly correlated to Aβ accumulation could be monitored, such as synaptic function, secretory vesicle function, and in the innate immune system.

5.3 Immunoprecipitation (IP) methods coupled with mass spectrometry

Due to wide dynamic range and low abundance of Aβ peptides, the most common experimental procedure to quantitate Aβ peptides in CSF or blood requires a sample preparation step before MS analysis. Many methods are currently available to purify/concentrate the Aβ peptides, such as solid-phase extraction (SPE), immunoprecipitation (IP), size exclusion, ultrafiltration and liquid-liquid extraction, immunodepletion, etc. [64]. Among these, IP is a common method. By using IP coupled with SRM-MS method, a recent publication reported that plasma $A\beta_{42}$ concentration correlated with the CSF $A\beta_{42}/A\beta_{40}$ ratio and had good accuracy for predicting the sensitivity and specificity of elevated brain Aβ [25]. Similarly, Nakamura et al. recently proposed a set of plasma biomarkers, the amyloid-β precursor protein (APP) 669–711/$A\beta_{42}$ and $A\beta_{40/42}$ ratios and their composites, for AD diagnosis with high sensitivity and specificity. Their composites displayed an accuracy of 90% in predicting Aβ brain burden at an individual level, as confirmed with PET imaging [84]. Of note, as reviewed by Brinkmalm et al. [85], in normal APP and Aβ metabolism, Aβ is most likely regulated by amyloid-degrading enzymes [86]. Different lengths of Aβ peptides exist in vivo, depending on different degradation pathways of APP [87]. To date, more than 40 different endogenous APP and Aβ peptides, including modifications, have been identified in the CSF [88]. Thus, these approaches can not only give a more accurate quantification of Aβ peptides in blood or CSF but also can be used to detect various Aβ species, which are beneficial to screen candidate biomarker for AD. For example, using the high selectivity of anti-Aβ antibodies in combination with mass spectrometry to determine the molecular mass with high accuracy, Vigo-Pelfrey et al. demonstrated the complex nature of Aβ peptides in the CSF and reported several different N- and C-terminal variants of Aβ [89]. In addition, IP-MS method has also been used to measure the protein levels in the CSF; using this method, a marked increase in the CSF levels of both synaptosomal-associated protein 25 (SNAP-25) and synaptotagmin-1 (SYT1) was found in AD dementia and prodromal AD cases [90, 91]. Interestingly, the levels of both SNAP-25 and SYT1 are reduced in cortical areas in the AD brain [90], thus suggesting that a set of synaptic proteins covering different components of the synaptic unit may be valuable tools in clinical studies on the relevance of synaptic dysfunction and degeneration in AD pathogenesis. This may also be used in the clinical evaluation of patients. The results indicate that this strategy is advantageous for detecting low abundance proteins, especially from CNS, or various Aβ peptides as a biomarker of AD.

6. Other technologies and methods

6.1 Metabolomics

Metabolomics is the newest omics platform that offers great potential for the diagnosis and prognosis of neurodegenerative diseases. This reflects alterations in genetics, transcription, and protein profiles and influences from the environment. Mass spectrometry (MS) and nuclear magnetic resonance (NMR) spectroscopy are two analytical platforms regularly used for detection. NMR is a particularly powerful tool for metabolite structural test. An MS-based approach is a sensitive one to identify and quantify in complex biological systems [65]. Metabolomics encompasses several techniques including untargeted metabolomics, targeted metabolomics, lipidomics, and fluxomics [92–94]. Untargeted metabolomics measures hundreds of metabolites in order to identify metabolic signatures related to a particular disease state or phenotype. This approach provides relative changes in metabolites and is useful for discovery projects where affected metabolic pathways are unknown. Targeted metabolomics provides quantitative measurements of a defined set of metabolites in a pathway of interest (e.g., glycolysis or TCA cycle). Lipidomics estimates changes in lipid profiles and requires specialized protocols for the detection and analysis of water-insoluble metabolites. Fluxomics incorporates stable isotope tracers to provide a dynamic, as opposed to static, assessment of metabolic changes and is performed in cells or in vivo [95]. Metabolomics has been widely used in the study of mechanisms and biomarkers of Alzheimer's disease. Metabolomics analysis conducted with biological samples of patients with MCI and AD identified metabolic changes associated with preclinical and clinical AD, such as plasma, CSF, and saliva (**Table 2**) [95–104]. These findings suggest that metabolomics-based biomarkers could be used to improve disease diagnosis, which will allow target pathways altered earlier in AD.

6.2 MicroRNA (miRNA)

miRNAs are a class of small non-coding RNAs of 20–22 nucleotides in length, which regulate more than 50% of protein-coding genes [105], and are associated with many neurodegenerative diseases, such as AD [106]. Fransquet and Ryan comprehensively reviewed the methods and findings from 26 studies comparing the measurement of miRNA in blood between AD cases and controls [107]. Of 8098 individually measured miRNAs, 23 that were differentially expressed miRNAs were found to be significant in two or more studies. Only six miRNAs (miR-107, miR-125b, miR-146a, miR-181c, miR-29b, and miR-342) were consistent in their direction of expression between these studies [107]. Interestingly, miR-107 has been found to be associated with the dysregulation of proteins involved in aspects of AD pathology, as well as being consistently downregulated in AD brains [107].Thus, the differentially expressed miRNAs and the corresponding targets will be potential biomarkers and provide evidence for new strategies for design of drugs for AD treatment.

6.3 Exosomes

Exosomes contain proteins, messenger RNAs (mRNAs), and microRNAs (miRNAs) that reflect their cellular origin, and they play a prominent role in cellular signaling, expulsion of toxic proteins, and transfer of cellular pathogens to other cells. CNS-derived exosomes (NEDs) are present in biological fluids (blood, CSF, and urine) and circulate in the interstitial space, both in the brain and in the periphery [108]. It may serve as markers of underlying CNS changes that occur in

Category	Samples	Methods	Candidate metabolites[a]	References
Untargeted metabolomic	Plasma	UPLC/HILIC -QTOF-MS	All groups: 4-aminobutanal ↓, spermine ↑ L-arginine ↑ L-ornithine ↑	[96]
Lipidomic	Plasma	UPLC-MS	Ceramides ↑: Cer16:0 Cer18:0, Cer24:1 Phosphatidylcholines ↓: PC36:5, PC38:6	[97]
Lipidomics	Plasma	UPLC-QTOF-MS	Phosphatidylcholine 40:4 ↑ Triglyceride 57:1 ↓ ChoE/triglyceride ↓	[98]
Targeted metabolomics	Serum	UPLC-TQ-S-MS/MS	Aβ pathology ↑: PCs and SMs Tau pathology ↑: long-chain acylcarnitines, PC ae C36:2, and SM.C20:2	[99]
Untargeted metabolomics	Serum	FIA-MS/MS UPLC-MS/MS	Glycerophospholipids ↓ Sphingolipids ↑	[100]
Untargeted metabolomic	Saliva	FUPLC-Q-TOF/MS	Phenyllactic acid ↑ hypoxanthine ↓ Sphinganine-1-phosphate ↑ Ornithine ↑, inosine ↓ 3-Dehydrocarnitine ↓	[101]
Targeted metabolomic	Saliva	^{1}H NMR	Propionates ↑	[102]
Untargeted metabolomics	CSF	UPLC-MS/MS	S-adenosylhomocysteine ↓, glycine ↓, S-adenosylmethionine ↑	[103]
Targeted metabolomics	PCSF	FMOC-derivatized UHPLC-MS/MS	Methionine sulfoxide ↑, guanine ↑, Anthranilate ↓, diacetylspermine ↓, 3-Methoxy-anthranilate ↑, Cadaverine ↑, histamine ↑, 3-HydroxyKynurenine ↓	[104]

Abbreviations: Cer: Ceramides; ChoE/TG: indicates co-elution of ChoE and TG molecules; FIA-MS/MS: flow injection analysis-MS/MS; FMOC: 9-fluorenylmethyl chloroformate; FUPLC: faster ultra-performance liquid chromatography; HILIC: hydrophilic interaction liquid chromatography; MS: mass spectrometry; NMR: nuclear magnetic resonance; PC: Phosphatidylcholines; PCSF: Postmortem cerebrospinal fluid; SM: sphingomyelin; TOF: Time of flight; UPLC: high-performance liquid chromatography.
[a]Compared with the control: ↑: upregulated; ↓: downregulated.

Table 2.
AD-related metabolomics studies reported in the literatures.

advance of changes in circulating proteins. Importantly, CNS-derived exosomes have unique surface markers that reflect their origin. By using the corresponding antibodies, targeted examinations of neuron-, astrocyte-, or endothelial cells can be performed (**Table 3**) [109–113]. Several proteins in neural-derived plasma exosomes have been identified to associate with preclinical AD [112], and cargo proteins of plasma astrocyte-derived exosomes in AD have also been detected [110]. Interestingly, alterations in plasma NDE levels of P-tau, $A\beta_{42}$, neurogranin, and repressor element 1-silencing transcription factor were found among AD and MCI cases that converted to AD within 36 months compared with stable MCI cases and normal control subjects [113]. In addition, miRNAs released from exosomes appear to be associated with multiple neurodegenerative conditions linking to AD, which is marked by hyperphosphorylated tau proteins and accumulation of Aβ plaques [114].

Samples	Materials	Exosomes-proteins/ miRNAs[a]	Methods	References
AD = 57 Control = 57	Plasma	Total tau ↑ P-T181-tau ↑ P-S396-tau ↑ Aβ1–42 ↑	ELISA	[109]
AD = 12 Control = 10	Plasma	GDNF ↓ P-T181-tau ↑ BACE-1 ↑ sAPPβ ↑ P-S396-tau ↑	ELISA	[110]
AD = 26 Control = 26	Plasma	P-serine-312-IRS-1 ↑	ELISA	[111]
AD = 46 Control = 46	Plasma	LAMP-1 ↑, Ubiquitin ↑, HSP70 ↓.	ELISA	[112]
AD = 10 MCI = 20 Control = 10	Plasma	P-T181-tau ↑ P-S396-tau ↑ Aβ1-42 ↑ NRGN ↓ REST ↓	ELISA	[113]
MCI = 43 DAT = 51	Plasma, CSF	mir-193b ↓	qPCR, WB	[115]
AD = 50 Control = 50	Plasma	mir-342-3p ↓ mir-342-5p ↓ mir-23b-3p ↓ mir-24-3p ↓ mir-338-3p ↓ mir-3065-5p ↓	MicroRNA sequencing	[116]
AD = 28 Control = 27	CSF	mir-29c ↓ mir-136-3p ↓ mir-16-2 ↓ mir-331-5p ↓ mir-485-5p ↑	miRNA assay qPCR	[117]
Health Control = 23/36 MCI = 3/8 AD = 23/16	Serum	Fold change > 1.5 Adjust p-value < 0.05 (HC vs AD): hsa-miR-20a-5p ↑ hsa-miR-3065-5pb ↑ hsa-miR-582-5p ↑ Fold change < 0.83 Adjust p-value < 0.05 (HC vs AD): hsa-miR-342-3p ↓ hsa-miR-1306-5p ↓	RT-qPCR Deep sequencing	[118]

Abbreviations: BACE-1: β-site amyloid precursor protein-cleaving enzyme 1; DAT: dementia of Alzheimer type; GDNF: glial-derived neurotrophic factor; HSP70: heat-shock protein 70; IRS: insulin receptor substrate; LAMP-1: lysosome-associated membrane protein 1; NRGN: neurogranin; REST: repressor element 1-silencing transcription factor; WB: Western blot analysis; RT-qPCR: Reverse transcription-quantitative real-time PCR.
[a]Expression changes in AD, compared with the control: ↑: upregulated; ↓: downregulated.

Table 3.
AD-related CNS-derived exosomes (proteins/miRNAs) reported in the literatures.

Specific profiles of exosomal miRNAs from human biological fluids, such as plasma and CSF, have prompted the potential application of miRNAs as diagnostic biomarkers (**Table 3**) [115–118]. These findings further support the search of exosome-based biomarkers for AD and other neurodegenerative diseases.

7. Conclusions

AD is the most common type of dementia and is becoming a major challenge for global health and social care. The last 20 years have seen an enormous expansion in research on biomarkers for AD. The use of biomarkers, such as T-tau, P-tau, and $A\beta_{42}$ (and $A\beta_{42}/A\beta_{40}$ ratio), together with brain imaging now provides the ability to detect evidence of the AD pathophysiological process in vivo. However, CSF biomarker and brain imaging are not used as screening tools. Research efforts have focused on the development and validation of non-invasive blood-based biomarkers. Recent advances in technical developments of novel ultrasensitive immunoassay, mass spectrometry methods, metabolomics, and exosomes show promise for blood biomarkers with potential applications as screening tools for AD (**Figure 1**). These opened a window for the study of AD biomarkers.

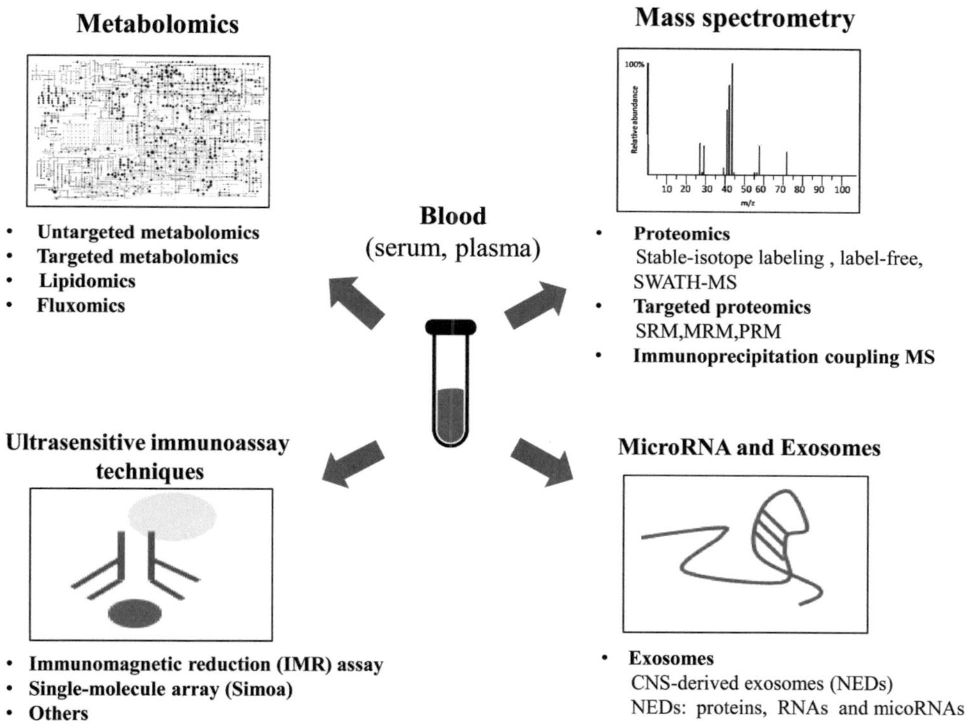

Figure 1.
Overview of the feasible and commonly used methods for AD blood biomarker discovery.

Acknowledgements

The authors would like to acknowledge the National Natural Science Foundation of China (Grant No. 31870825, 21877081) and Shenzhen Bureau of Science, Technology and Information (Nos. JCYJ20150402100258220, JCYJ20150529164656093, JCYJ20170412110026229) for funds to support this work.

Conflict of interest

The authors declare that they have no competing interests.

Author details

Liming Shen*, Sijian Xia, Huajie Zhang, Fang Yao, Xukun Liu, Yuxi Zhao, Ming Ying, Javed Iqbal and Qiong Liu
College of Life Science and Oceanography, Shenzhen University, Shenzhen, P.R. China

*Address all correspondence to: slm@szu.edu.cn

References

[1] Zhang Z-H, Wu Q-Y, Zheng R, Chen C, Chen Y, Liu Q, et al. Selenomethionine mitigates cognitive decline by targeting both tau hyperphosphorylation and autophagic clearance in an Alzheimer's disease mouse model. The Journal of Neuroscience. 2017;**37**:2449-2462. DOI: 10.1523/JNEUROSCI.3229-16.2017

[2] Blennow K, de Leon MJ, Zetterberg H. Alzheimer's disease. Lancet. 2006;**368**:387-403. DOI: 10.1016/S0140-6736(06)69113-7

[3] Alzheimer's Association. 2010 Alzheimer's disease facts and figures. Alzheimer's & Dementia. 2010;**6**:158-194. DOI: 10.1016/j.jalz.2010.01.009

[4] Alzheimer's Association. 2018 Alzheimer's disease facts and figures. Alzheimer's & Dementia: The Journal of the Alzheimer's Association. 2018;**14**:367-429. DOI: 10.1016/j.jalz.2018.02.001

[5] Prince M, Ali G-C, Guerchet M, Prina AM, Albanese E, Wu Y-T. Recent global trends in the prevalence and incidence of dementia, and survival with dementia. Alzheimer's Research & Therapy. 2016;**8**:23. DOI: 10.1186/s13195-016-0188-8

[6] Feng L, Li J, Yu J-T, Zhang C, Yang B, Vellas B, et al. Editorial: Prevention of Alzheimer's disease in Chinese populations: Status, challenges and directions. The Journal of Prevention of Alzheimer's Disease. 2018;**5**:90-94. DOI: 10.14283/jpad.2018.14

[7] Freudenberg-Hua Y, Li W, Davies P. The role of genetics in advancing precision medicine for Alzheimer's disease—A narrative review. Frontiers in Medicine (Lausanne). 2018;**5**:108. DOI: 10.3389/fmed.2018.00108

[8] Rajasekhar K, Govindaraju T. Current progress, challenges and future prospects of diagnostic and therapeutic interventions in Alzheimer's disease. RSC Advances. 2018;**8**:23780-23804. DOI: 10.1039/C8RA03620A

[9] Schon EA, Area-Gomez E. Mitochondria-associated ER membranes in Alzheimer disease. Molecular and Cellular Neurosciences. 2013;**55**:26-36. DOI: 10.1016/j.mcn.2012.07.011

[10] Jivad N, Rabiei Z. A review study on medicinal plants used in the treatment of learning and memory impairments. Asian Pacific Journal of Tropical Biomedicine. 2014;**4**:780-789. DOI: 10.12980/APJTB.4.2014APJTB-2014-0412

[11] Sabbagh MN, Lue L-F, Fayard D, Shi J. Increasing precision of clinical diagnosis of Alzheimer's disease using a combined algorithm incorporating clinical and novel biomarker data. Neurology and Therapy. 2017;**6**:83-95. DOI: 10.1007/s40120-017-0069-5

[12] disease RAA. The search for a blood-based biomarker for Alzheimer disease. Nature Reviews. Neurology. 2014;**10**:618-619. DOI: 10.1038/nrneurol.2014.182

[13] Biagioni MC, Galvin JE. Using biomarkers to improve detection of Alzheimer's disease. Neurodegenerative Disease Management. 2011;**1**:127-139. DOI: 10.2217/NMT.11.11

[14] Ritchie K, Ritchie CW, Yaffe K, Skoog I, Scarmeas N. Is late-onset Alzheimer's disease really a disease of midlife? Alzheimer's & Dementia (NY). 2015;**1**:122-130. DOI: 10.1016/j.trci.2015.06.004

[15] McKhann G, Drachman D, Folstein M, Katzman R, Price D, Stadlan EM. Clinical diagnosis of Alzheimer's disease: report of the

NINCDS-ADRDA Work Group under the auspices of Department of Health and Human Services Task Force on Alzheimer's disease. Neurology. 1984;**34**:939-944. DOI: 10.1212/01.wnl.0000400650.92875.cf

[16] Dubois B, Feldman HH, Jacova C, Hampel H, Molinuevo JL, Blennow K, et al. Advancing research diagnostic criteria for Alzheimer's disease: The IWG-2 criteria. Lancet Neurology. 2014;**13**:614-629. DOI: 10.1016/S1474-4422(14)70090-0

[17] Dubois B, Feldman HH, Jacova C, Dekosky ST, Barberger-Gateau P, Cummings J, et al. Research criteria for the diagnosis of Alzheimer's disease: Revising the NINCDS-ADRDA criteria. Lancet Neurology. 2007;**6**:734-746. DOI: 10.1016/S1474-4422(07)70178-3

[18] Sperling RA, Aisen PS, Beckett LA, Bennett DA, Craft S, Fagan AM, et al. Toward defining the preclinical stages of Alzheimer's disease: Recommendations from the National Institute on Aging-Alzheimer's Association workgroups on diagnostic guidelines for Alzheimer's disease. Alzheimer's & Dementia. 2011;**7**:280-292. DOI: 10.1016/j.jalz.2011.03.003

[19] McKhann GM, Knopman DS, Chertkow H, Hyman BT, Jack CR, Kawas CH, et al. The diagnosis of dementia due to Alzheimer's disease: Recommendations from the National Institute on Aging-Alzheimer's Association workgroups on diagnostic guidelines for Alzheimer's disease. Alzheimer's & Dementia. 2011;**7**:263-269. DOI: 10.1016/j.jalz.2011.03.005

[20] Albert MS, DeKosky ST, Dickson D, Dubois B, Feldman HH, Fox NC, et al. The diagnosis of mild cognitive impairment due to Alzheimer's disease: Recommendations from the National Institute on Aging-Alzheimer's Association workgroups on diagnostic guidelines for Alzheimer's disease. Alzheimer's & Dementia. 2011;**7**:270-279. DOI: 10.1016/j.jalz.2011.03.008

[21] Jack CR, Albert MS, Knopman DS, McKhann GM, Sperling RA, Carrillo MC, et al. Introduction to the recommendations from the National Institute on Aging-Alzheimer's Association workgroups on diagnostic guidelines for Alzheimer's disease. Alzheimer's & Dementia. 2011;**7**:257-262. DOI: 10.1016/j.jalz.2011.03.004

[22] Hyman BT, Phelps CH, Beach TG, Bigio EH, Cairns NJ, Carrillo MC, et al. National Institute on Aging-Alzheimer's Association guidelines for the neuropathologic assessment of Alzheimer's disease. Alzheimer's & Dementia. 2012;**8**:1-13. DOI: 10.1016/j.jalz.2011.10.007

[23] Jack CR, Knopman DS, Jagust WJ, Shaw LM, Aisen PS, Weiner MW, et al. Hypothetical model of dynamic biomarkers of the Alzheimer's pathological cascade. Lancet Neurology. 2010;**9**:119-128. DOI: 10.1016/S1474-4422(09)70299-6

[24] Jack CR, Knopman DS, Jagust WJ, Petersen RC, Weiner MW, Aisen PS, et al. Update on hypothetical model of Alzheimer's disease biomarkers. Lancet Neurology. 2013;**12**:207-216. DOI: 10.1016/S1474-4422(12)70291-0

[25] Jack CR, Bennett DA, Blennow K, Carrillo MC, Dunn B, Haeberlein SB, et al. NIA-AA Research Framework: Toward a biological definition of Alzheimer's disease. Alzheimer's & Dementia. 2018;**14**:535-562. DOI: 10.1016/j.jalz.2018.02.018

[26] Lanoiselée H-M, Nicolas G, Wallon D, Rovelet-Lecrux A, Lacour M, Rousseau S, et al. APP, PSEN1, and PSEN2 mutations in early-onset Alzheimer disease: A genetic screening study of familial and sporadic cases. PLoS Medicine. 2017;**14**:e1002270. DOI: 10.1371/journal.pmed.1002270

[27] Barber RC. The genetics of Alzheimer's disease. Scientifica (Cairo). 2012;**2012**:246210. DOI: 10.6064/2012/246210

[28] Zou Z, Liu C, Che C, Huang H. Clinical genetics of Alzheimer's disease. BioMed Research International. 2014;**2014**:291862. DOI: 10.1155/2014/291862

[29] Guerreiro RJ, Gustafson DR, Hardy J. The genetic architecture of Alzheimer's disease: Beyond APP, PSENs and APOE. Neurobiology of Aging. 2012;**33**:437-456. DOI: 10.1016/j.neurobiolaging.2010.03.025

[30] Lista S, O'Bryant SE, Blennow K, Dubois B, Hugon J, Zetterberg H, et al. Biomarkers in sporadic and familial Alzheimer's disease. Journal of Alzheimer's Disease. 2015;**47**:291-317. DOI: 10.3233/JAD-143006

[31] Jack CR, Bennett DA, Blennow K, Carrillo MC, Feldman HH, Frisoni GB, et al. A/T/N: An unbiased descriptive classification scheme for Alzheimer disease biomarkers. Neurology. 2016;**87**:539-547. DOI: 10.1212/WNL.0000000000002923

[32] Herholz K, Ebmeier K. Clinical amyloid imaging in Alzheimer's disease. Lancet Neurology. 2011;**10**:667-670. DOI: 10.1016/S1474-4422(11)70123-5

[33] Olsson B, Lautner R, Andreasson U, Öhrfelt A, Portelius E, Bjerke M, et al. CSF and blood biomarkers for the diagnosis of Alzheimer's disease: A systematic review and meta-analysis. Lancet Neurology. 2016;**15**:673-684. DOI: 10.1016/S1474-4422(16)00070-3

[34] Blennow K, Zetterberg H, Fagan AM. Fluid biomarkers in Alzheimer disease. Cold Spring Harbor Perspectives in Medicine. 2012;**2**:a006221. DOI: 10.1101/cshperspect.a006221

[35] El Kadmiri N, Said N, Slassi I, El Moutawakil B, Nadifi S. Biomarkers for Alzheimer Disease: Classical and Novel Candidates' Review. Neuroscience 2018;**370**:181-190. DOI: 10.1016/j.neuroscience.2017.07.017

[36] Berti V, Polito C, Lombardi G, Ferrari C, Sorbi S, Pupi A. Rethinking on the concept of biomarkers in preclinical Alzheimer's disease. Neurological Sciences. 2016;**37**:663-672. DOI: 10.1007/s10072-016-2477-1

[37] Leinenbach A, Pannee J, Dülffer T, Huber A, Bittner T, Andreasson U, et al. Mass spectrometry-based candidate reference measurement procedure for quantification of amyloid-β in cerebrospinal fluid. Clinical Chemistry. 2014;**60**:987-994. DOI: 10.1373/clinchem.2013.220392

[38] Kuhlmann J, Andreasson U, Pannee J, Bjerke M, Portelius E, Leinenbach A, et al. CSF $A\beta_{1-42}$—An excellent but complicated Alzheimer's biomarker—A route to standardisation. Clinica Chimica Acta. 2017;**467**:27-33. DOI: 10.1016/j.cca.2016.05.014

[39] Kvartsberg H, Duits FH, Ingelsson M, Andreasen N, Öhrfelt A, Andersson K, et al. Cerebrospinal fluid levels of the synaptic protein neurogranin correlates with cognitive decline in prodromal Alzheimer's disease. Alzheimer's & Dementia. 2015;**11**:1180-1190. DOI: 10.1016/j.jalz.2014.10.009

[40] Ramesh S, Govindarajulu M, Jones E, Knowlton S, Weeks L, Suppiramaniam V, et al. Current and Novel Biomarkers for Alzheimer's Disease. 2018. Available from: https://meddocsonline.org/ebooks/alzheimers-disease-and-treatmen/current-and-novel-biomarkers-for-alzheimers-disease.pdf

[41] Blennow K, Zetterberg H. The past and the future of Alzheimer's disease fluid biomarkers. Journal of Alzheimer's Disease. 2018;**62**:1125-1240. DOI: 10.3233/JAD-170773

[42] Counts SE, Ikonomovic MD, Mercado N, Vega IE, Mufson EJ. Biomarkers for the early detection and progression of Alzheimer's disease. Neurotherapeutics. 2017;**14**:35-53. DOI: 10.1007/s13311-016-0481-z

[43] Aizenstein HJ, Nebes RD, Saxton JA, Price JC, Mathis CA, Tsopelas ND, et al. Frequent amyloid deposition without significant cognitive impairment among the elderly. Archives of Neurology. 2008;**65**:1509-1517. DOI: 10.1001/archneur.65.11.1509

[44] Roe CM, Fagan AM, Williams MM, Ghoshal N, Aeschleman M, Grant EA, et al. Improving CSF biomarker accuracy in predicting prevalent and incident Alzheimer disease. Neurology. 2011;**76**:501-510. DOI: 10.1212/WNL.0b013e31820af900

[45] Hampel H, Toschi N, Baldacci F, Zetterberg H, Blennow K, Kilimann I, et al. Alzheimer's disease biomarker-guided diagnostic workflow using the added value of six combined cerebrospinal fluid candidates: $A\beta_{1-42}$, total-tau, phosphorylated-tau, NFL, neurogranin, and YKL-40. Alzheimer's & Dementia. 2018;**14**:492-501. DOI: 10.1016/j.jalz.2017.11.015

[46] Mattsson N, Andreasson U, Zetterberg H, Blennow K. Alzheimer's disease neuroimaging initiative, Association of plasma neurofilament light with neurodegeneration in patients with Alzheimer disease. JAMA Neurology. 2017;**74**:557-566. DOI: 10.1001/jamaneurol.2016.6117

[47] Roberts KF, Elbert DL, Kasten TP, Patterson BW, Sigurdson WC, Connors RE, et al. Amyloid-β efflux from the central nervous system into the plasma. Annals of Neurology. 2014;**76**:837-844. DOI: 10.1002/ana.24270

[48] Hanon O, Vidal J-S, Lehmann S, Bombois S, Allinquant B, Tréluyer J-M, et al. Plasma amyloid levels within the Alzheimer's process and correlations with central biomarkers. Alzheimer's & Dementia. 2018;**14**:858-868. DOI: 10.1016/j.jalz.2018.01.004

[49] Koyama A, Okereke OI, Yang T, Blacker D, Selkoe DJ, Grodstein F. Plasma amyloid-β as a predictor of dementia and cognitive decline: a systematic review and meta-analysis. Archives of Neurology. 2012;**69**:824-831. DOI: 10.1001/archneurol.2011.1841

[50] Andreasson U, Blennow K, Zetterberg H. Update on ultrasensitive technologies to facilitate research on blood biomarkers for central nervous system disorders. Alzheimer's & Dementia (Amst). 2016;**3**:98-102. DOI: 10.1016/j.dadm.2016.05.005

[51] Rissin DM, Kan CW, Campbell TG, Howes SC, Fournier DR, Song L, et al. Single-molecule enzyme-linked immunosorbent assay detects serum proteins at subfemtomolar concentrations. Nature Biotechnology. 2010;**28**:595-599. DOI: 10.1038/nbt.1641

[52] Wilson DH, Rissin DM, Kan CW, Fournier DR, Piech T, Campbell TG, et al. The Simoa HD-1 Analyzer: A novel fully automated digital immunoassay analyzer with single-molecule sensitivity and multiplexing. Journal of Laboratory Automation. 2016;**21**:533-547. DOI: 10.1177/2211068215589580

[53] Mattsson N, Zetterberg H, Janelidze S, Insel PS, Andreasson U, Stomrud E, et al. Plasma tau in Alzheimer disease. Neurology. 2016;**87**:1827-1835. DOI: 10.1212/WNL.0000000000003246

[54] Tatebe H, Kasai T, Ohmichi T, Kishi Y, Kakeya T, Waragai M, et al. Quantification of plasma phosphorylated tau to use as a biomarker for brain Alzheimer pathology: pilot case-control studies including patients with Alzheimer's disease and down syndrome. Molecular

Neurodegeneration. 2017;**12**:63. DOI: 10.1186/s13024-017-0206-8

[55] Mielke MM, Hagen CE, Xu J, Chai X, Vemuri P, Lowe VJ, et al. Plasma phospho-tau181 increases with Alzheimer's disease clinical severity and is associated with tau- and amyloid-positron emission tomography. Alzheimer's & Dementia. 2018;**14**: 989-997. DOI: 10.1016/j.jalz.2018.02.013

[56] Chieh JJ, Yang SY, Jian ZF, Wang WC, Hornga HE, Yang HC, et al. Hyper-high-sensitivity wash-free magnetoreduction assay on biomolecules using high-T-c superconducting quantum interference devices. Journal of Applied Physics. 2008;**103**:1-6. DOI: 10.1063/1.2827359

[57] Yang C-C, Yang S-Y, Chieh J-J, Horng H-E, Hong C-Y, Yang H-C, et al. Biofunctionalized magnetic nanoparticles for specifically detecting biomarkers of Alzheimer's disease in vitro. ACS Chemical Neuroscience. 2011;**2**:500-505. DOI: 10.1021/cn200028j

[58] Chiu MJ, Yang SY, Chen TF, Chieh JJ, Huang TZ, Yip PK, et al. New assay for old markers-plasma beta amyloid of mild cognitive impairment and Alzheimer's disease. Current Alzheimer Research. 2012;**9**:1142-1148. DOI: 10.2174/156720512804142967

[59] Teunissen CE, Chiu M-J, Yang C-C, Yang S-Y, Scheltens P, Zetterberg H, et al. Plasma Amyloid-β ($A\beta_{42}$) correlates with cerebrospinal fluid $A\beta_{42}$ in Alzheimer's disease. Journal of Alzheimer's Disease. 2018;**62**:1857-1863. DOI: 10.3233/JAD-170784

[60] Lue L-F, Sabbagh MN, Chiu M-J, Jing N, Snyder NL, Schmitz C, et al. Plasma levels of $A\beta_{42}$ and tau identified probable Alzheimer's dementia: Findings in two cohorts. Frontiers in Aging Neuroscience. 2017;**9**:226. DOI: 10.3389/fnagi.2017.00226

[61] Yang C-C, Chiu M-J, Chen T-F, Chang H-L, Liu B-H, Yang S-Y. Assay of plasma phosphorylated tau protein (Threonine 181) and total tau protein in early-stage Alzheimer's disease. Journal of Alzheimer's Disease. 2018;**61**:1323-1332. DOI: 10.3233/JAD-170810

[62] Snyder HM, Carrillo MC, Grodstein F, Henriksen K, Jeromin A, Lovestone S, et al. Developing novel blood-based biomarkers for Alzheimer's disease. Alzheimer's & Dementia. 2014;**10**: 109-114. DOI: 10.1016/j.jalz.2013.10.007

[63] Robinson RAS, Amin B, Guest PC. Multiplexing biomarker methods, proteomics and considerations for Alzheimer's disease. Advances in Experimental Medicine and Biology. 2017;**974**:21-48. DOI: 10.1007/978-3-319-52479-5_2

[64] Grasso G. Mass spectrometry is a multifaceted weapon to be used in the battle against Alzheimer's disease: Amyloid beta peptides and beyond. Mass Spectrometry Reviews 2018.**7**:1-15. DOI: 10.10-02/mas.21566

[65] Liu Y, Qing H, Deng Y. Biomarkers in Alzheimer's disease analysis by mass spectrometry-based proteomics. International Journal of Molecular Sciences. 2014;**15**:7865-7882. DOI: 10.3390/ijms15057865

[66] Paul D, Kumar A, Gajbhiye A, Santra MK, Srikanth R. Mass spectrometry-based proteomics in molecular diagnostics: discovery of cancer biomarkers using tissue culture. BioMed Research International. 2013;**2013**:783131. DOI: 10.1155/2013/783131

[67] Shen L, Liao L, Chen C, Guo Y, Song D, Wang Y, et al. Proteomics analysis of blood serums from Alzheimer's disease patients using iTRAQ labeling technology. Journal of Alzheimer's Disease. 2017;**56**:361-378. DOI: 10.3233/JAD-160913

[68] Song F, Poljak A, Kochan NA, Raftery M, Brodaty H, Smythe GA, et al. Plasma protein profiling of mild cognitive impairment and Alzheimer's disease using iTRAQ quantitative proteomics. Proteome Science. 2014;**12**:5. DOI: 10.1186/1477-5956-12-5

[69] Muenchhoff J, Poljak A, Song F, Raftery M, Brodaty H, Duncan M, et al. Plasma protein profiling of mild cognitive impairment and Alzheimer's disease across two independent cohorts. Journal of Alzheimer's Disease. 2015;**43**:1355-1373. DOI: 10.3233/JAD-141266

[70] Ludwig C, Gillet L, Rosenberger G, Amon S, Collins BC, Aebersold R. Data-independent acquisition-based SWATH-MS for quantitative proteomics: A tutorial. Molecular Systems Biology. 2018;**14**:e8126. DOI: 10.15252/msb.20178126

[71] Zhang K, Schrag M, Crofton A, Trivedi R, Vinters H, Kirsch W. Targeted proteomics for quantification of histone acetylation in Alzheimer's disease. Proteomics. 2012;**12**:1261-1268. DOI: 10.1002/pmic.201200010

[72] Schrag M, Mueller C, Zabel M, Crofton A, Kirsch WM, Ghribi O, et al. Oxidative stress in blood in Alzheimer's disease and mild cognitive impairment: A meta-analysis. Neurobiology of Disease. 2013;**59**:100-110. DOI: 10.1016/j.nbd.2013.07.005

[73] Shen L, Chen C, Yang A, Chen Y, Liu Q, Ni J. Redox proteomics identification of specifically carbonylated proteins in the hippocampi of triple transgenic Alzheimer's disease mice at its earliest pathological stage. Journal of Proteomics. 2015;**123**:101-113. DOI: 10.1016/j.jprot.2015.04.005

[74] Shen L, Chen Y, Yang A, Chen C, Liao L, Li S, et al. Redox proteomic profiling of specifically carbonylated proteins in the serum of triple transgenic Alzheimer's disease mice. International Journal of Molecular Sciences. 2016;**17**:469. DOI: 10.3390/ijms17040469

[75] Baldeiras I, Santana I, Proença MT, Garrucho MH, Pascoal R, Rodrigues A, et al. Peripheral oxidative damage in mild cognitive impairment and mild Alzheimer's disease. Journal of Alzheimer's Disease. 2008;**15**:117-128. DOI: 10.3233/JAD-2008-15110

[76] Zafrilla P, Mulero J, Xandri JM, Santo E, Caravaca G, Morillas JM. Oxidative stress in Alzheimer patients in different stages of the disease. Current Medicinal Chemistry. 2006;**13**:1075-1083. DOI: 10.2174/092986706776360978

[77] Harlan R, Zhang H. Targeted proteomics: A bridge between discovery and validation. Expert Review of Proteomics. 2014;**11**:657-661. DOI: 10.1586/14789450.2014.976558

[78] Chambers AG, Percy AJ, Simon R, Borchers CH. MRM for the verification of cancer biomarker proteins: recent applications to human plasma and serum. Expert Review of Proteomics. 2014;**11**:137-148. DOI: 10.1586/14789450.2014.877346

[79] Pannee J, Portelius E, Oppermann M, Atkins A, Hornshaw M, Zegers I, et al. A selected reaction monitoring (SRM)-based method for absolute quantification of $A\beta_{38}$, $A\beta_{40}$, and $A\beta_{42}$ in cerebrospinal fluid of Alzheimer's disease patients and healthy controls. Journal of Alzheimer's Disease. 2013;**33**:1021-1032. DOI: 10.3233/JAD-2012-121471

[80] Ringman JM, Schulman H, Becker C, Jones T, Bai Y, Immermann F, et al. Proteomic changes in cerebrospinal fluid of presymptomatic and affected

persons carrying familial Alzheimer disease mutations. Archives of Neurology. 2012;**69**:96-104. DOI: 10.1001/archneurol.2011.642

[81] Wildsmith KR, Schauer SP, Smith AM, Arnott D, Zhu Y, Haznedar J, et al. Identification of longitudinally dynamic biomarkers in Alzheimer's disease cerebrospinal fluid by targeted proteomics. Molecular Neurodegeneration. 2014;**9**:22. DOI: 10.1186/1750-1326-9-22

[82] Brinkmalm G, Sjödin S, Simonsen AH, Hasselbalch SG, Zetterberg H, Brinkmalm A, et al. A Parallel Reaction Monitoring Mass Spectrometric Method for Analysis of Potential CSF Biomarkers for Alzheimer's Disease. Proteomics Clinical Applications. 2018;**12**:1-13. DOI: 10.1002/prca.201700131

[83] Duits FH, Brinkmalm G, Teunissen CE, Brinkmalm A, Scheltens P, Van der Flier WM, et al. Synaptic proteins in CSF as potential novel biomarkers for prognosis in prodromal Alzheimer's disease. Alzheimer's Research & Therapy. 2018;**10**:15. DOI: 10.1186/s13195-017-0335-x

[84] Nakamura A, Kaneko N, Villemagne VL, Kato T, Doecke J, Doré V, et al. High performance plasma amyloid-β biomarkers for Alzheimer's disease. Nature. 2018;**554**:249-254. DOI: 10.1038/nature25456

[85] Brinkmalm A, Portelius E, Öhrfelt A, Brinkmalm G, Andreasson U, Gobom J, et al. Explorative and targeted neuroproteomics in Alzheimer's disease. Biochimica et Biophysica Acta. 2015;**1854**:769-778. DOI: 10.1016/j.bbapap.2015.01.009

[86] Nalivaeva NN, Beckett C, Belyaev ND, Turner AJ. Are amyloid-degrading enzymes viable therapeutic targets in Alzheimer's disease? Journal of Neurochemistry. 2012;**120**(Suppl 1):167-185. DOI: 10.1111/j.1471-4159.2011.07510.x

[87] Portelius E, Mattsson N, Andreasson U, Blennow K, Zetterberg H. Novel aβ isoforms in Alzheimer's disease—Their role in diagnosis and treatment. Current Pharmaceutical Design. 2011;**17**:2594-2602. DOI: 10.2174/138161211797416039

[88] Brinkmalm G, Portelius E, Öhrfelt A, Mattsson N, Persson R, Gustavsson MK, et al. An online nano-LC-ESI-FTICR-MS method for comprehensive characterization of endogenous fragments from amyloid β and amyloid precursor protein in human and cat cerebrospinal fluid. Journal of Mass Spectrometry. 2012;**47**:591-603. DOI: 10.1002/jms.2987

[89] Vigo-Pelfrey C, Lee D, Keim P, Lieberburg I, Schenk DB. Characterization of beta-amyloid peptide from human cerebrospinal fluid. Journal of Neurochemistry. 1993;**61**:1965-1968

[90] Brinkmalm A, Brinkmalm G, Honer WG, Frölich L, Hausner L, Minthon L, et al. SNAP-25 is a promising novel cerebrospinal fluid biomarker for synapse degeneration in Alzheimer's disease. Molecular Neurodegeneration. 2014;**9**:53. DOI: 10.1186/1750-1326-9-53

[91] Öhrfelt A, Brinkmalm A, Dumurgier J, Brinkmalm G, Hansson O, Zetterberg H, et al. The pre-synaptic vesicle protein synaptotagmin is a novel biomarker for Alzheimer's disease. Alzheimer's Research & Therapy. 2016;**8**:41. DOI: 10.1186/s13195-016-0208-8

[92] Patti GJ, Yanes O, Siuzdak G. Innovation: Metabolomics: The apogee of the omics trilogy. Nature Reviews. Molecular Cell Biology. 2012;**13**:263-269. DOI: 10.1038/nrm3314

[93] Charidemou E, Ashmore T, Griffin JL. The use of stable isotopes in the study of human pathophysiology. The International Journal of Biochemistry & Cell Biology. 2017;**93**:102-109. DOI: 10.1016/j.biocel.2017.07.012

[94] Triebl A, Trötzmüller M, Hartler J, Stojakovic T, Köfeler HC. Lipidomics by ultrahigh performance liquid chromatography-high resolution mass spectrometry and its application to complex biological samples. Journal of Chromatography. B, Analytical Technologies in the Biomedical and Life Sciences. 2017;**1053**:1072-1080. DOI: 10.1016/j.jchromb.2017.03.027

[95] Wilkins JM, Trushina E. Application of metabolomics in Alzheimer's disease. Frontiers in Neurology. 2017;**8**:719. DOI: 10.3389/fneur.2017.00719

[96] Graham SF, Chevallier OP, Elliott CT, Hölscher C, Johnston J, McGuinness B, et al. Untargeted Metabolomic analysis of human plasma indicates differentially affected polyamine and L-Arginine metabolism in mild cognitive impairment subjects converting to Alzheimer's disease. PLoS One. 2015;**10**:e0119452. DOI: 10.1371/journal.pone.0119452

[97] Kim M, Nevado-Holgado A, Whiley L, Snowden SG, Soininen H, Kloszewska I, et al. Association between plasma ceramides and phosphatidylcholines and hippocampal brain volume in late onset Alzheimer's disease. Journal of Alzheimer's Disease. 2017;**60**:809-817. DOI: 10.3233/JAD-160645

[98] Proitsi P, Kim M, Whiley L, Simmons A, Sattlecker M, Velayudhan L, et al. Association of blood lipids with Alzheimer's disease: A comprehensive lipidomics analysis. Alzheimer's & Dementia. 2017;**13**:140-151. DOI: 10.1016/j.jalz.2016.08.003

[99] Toledo JB, Arnold M, Kastenmüller G, Chang R, Baillie RA, Han X, et al. Metabolic network failures in Alzheimer's disease: A biochemical road map. Alzheimer's & Dementia. 2017;**13**:965-984. DOI: 10.1016/j.jalz.2017.01.020

[100] Varma VR, Oommen AM, Varma S, Casanova R, An Y, Andrews RM, et al. Brain and blood metabolite signatures of pathology and progression in Alzheimer disease: A targeted metabolomics study. PLoS Medicine. 2018;**15**:e1002482. DOI: 10.1371/journal.pmed.1002482

[101] Liang Q, Liu H, Zhang T, Jiang Y, Xing H, Zhang A. Metabolomics-based screening of salivary biomarkers for early diagnosis of Alzheimer's disease. RSC Advances. 2015;**5**:96074-96079. DOI: 10.1039/c5ra19094k

[102] Yilmaz A, Geddes T, Han B, Bahado-Singh RO, Wilson GD, Imam K, et al. Diagnostic biomarkers of Alzheimer's disease as identified in saliva using 1H NMR-based metabolomics. Journal of Alzheimer's Disease. 2017;**58**:355-359. DOI: 10.3233/JAD-161226

[103] Guiraud SP, Montoliu I, Da Silva L, Dayon L, Galindo AN, Corthésy J, et al. High-throughput and simultaneous quantitative analysis of homocysteine–methionine cycle metabolites and co-factors in blood plasma and cerebrospinal fluid by isotope dilution LC–MS/MS. Analytical and Bioanalytical Chemistry. 2017;**409**:295-305. DOI: 10.1007/s00216-016-0003-1

[104] Muguruma Y, Tsutsui H, Noda T, Akatsu H, Inoue K. Widely targeted metabolomics of Alzheimer's disease postmortem cerebrospinal fluid based on 9-fluorenylmethyl chloroformate derivatized ultra-high performance liquid chromatography tandem mass spectrometry. Journal of

Chromatography B. 2018;**1091**:53-66. DOI: 10.1016/j.jchromb.2018.05.031

[105] Maffioletti E, Tardito D, Gennarelli M, Bocchio-Chiavetto L. Micro spies from the brain to the periphery: New clues from studies on microRNAs in neuropsychiatric disorders. Frontiers in Cellular Neuroscience. 2014;**8**:75. DOI: 10.3389/fncel.2014.00075

[106] Grasso M, Piscopo P, Confaloni A, Denti MA. Circulating miRNAs as biomarkers for neurodegenerative disorders. Molecules. 2014;**19**:6891-6910. DOI: 10.3390/molecules19056891

[107] Fransquet PD, Ryan J. Micro RNA as a potential blood-based epigenetic biomarker for Alzheimer's disease. Clinical Biochemistry. 2018;**58**:5-14. DOI: 10.1016/j.clinbiochem.2018.05.020

[108] Barile L, Vassalli G. Exosomes: Therapy delivery tools and biomarkers of diseases. Pharmacology & Therapeutics. 2017;**174**:63-78. DOI: 10.1016/j.pharmthera.2017.02.020

[109] Fiandaca MS, Kapogiannis D, Mapstone M, Boxer A, Eitan E, Schwartz JB, et al. Identification of preclinical Alzheimer's disease by a profile of pathogenic proteins in neurally derived blood exosomes: A case-control study. Alzheimer's & Dementia. 2015;**11**:600-607.e1. DOI: 10.1016/j.jalz.2014.06.008

[110] Goetzl EJ, Mustapic M, Kapogiannis D, Eitan E, Lobach IV, Goetzl L, et al. Cargo proteins of plasma astrocyte-derived exosomes in Alzheimer's disease. The FASEB Journal. 2016;**30**:3853-3859. DOI: 10.1096/fj.201600756R

[111] Kapogiannis D, Boxer A, Schwartz JB, Abner EL, Biragyn A, Masharani U, et al. Dysfunctionally phosphorylated type 1 insulin receptor substrate in neural-derived blood exosomes of preclinical Alzheimer's disease. The FASEB Journal. 2015;**29**:589-596. DOI: 10.1096/fj.14-262048

[112] Goetzl EJ, Boxer A, Schwartz JB, Abner EL, Petersen RC, Miller BL, et al. Altered lysosomal proteins in neural-derived plasma exosomes in preclinical Alzheimer disease. Neurology. 2015;**85**:40-47. DOI: 10.1212/WNL.0000000000001702

[113] Winston CN, Goetzl EJ, Akers JC, Carter BS, Rockenstein EM, Galasko D, et al. Prediction of conversion from mild cognitive impairment to dementia with neuronally derived blood exosome protein profile. Alzheimer's & Dementia (Amst). 2016;**3**:63-72. DOI: 10.1016/j.dadm.2016.04.001

[114] Chen J-J, Zhao B, Zhao J, Li S. Potential roles of exosomal microRNAs as diagnostic biomarkers and therapeutic application in Alzheimer's disease. Neural Plasticity. 2017;**2017**:7027380. DOI: 10.1155/2017/7027380

[115] Liu C-G, Song J, Zhang Y-Q, Wang P-C. MicroRNA-193b is a regulator of amyloid precursor protein in the blood and cerebrospinal fluid derived exosomal microRNA-193b is a biomarker of Alzheimer's disease. Molecular Medicine Reports. 2014;**10**:2395-2400. DOI: 10.3892/mmr.2014.2484

[116] Lugli G, Cohen AM, Bennett DA, Shah RC, Fields CJ, Hernandez AG, et al. Plasma exosomal miRNAs in persons with and without Alzheimer disease: Altered expression and prospects for biomarkers. PLoS One. 2015;**10**:e0139233. DOI: 10.1371/journal.pone.0139233

[117] Gui Y, Liu H, Zhang L, Lv W, Hu X. Altered microRNA profiles

in cerebrospinal fluid exosome in Parkinson disease and Alzheimer disease. Oncotarget. 2015;**6**:37043-37053. DOI: 10.18632/oncotarget.6158

[118] Cheng L, Doecke JD, Sharples RA, Villemagne VL, Fowler CJ, Rembach A, et al. Prognostic serum miRNA biomarkers associated with Alzheimer's disease shows concordance with neuropsychological and neuroimaging assessment. Molecular Psychiatry. 2015;**20**:1188-1196. DOI: 10.1038/mp.2014.127

Chapter 3

Pharmacogenetics and Cancer Treatment: Progress and Prospects

Munindra Ruwali

Abstract

The response of cancer patients to chemotherapy follows a very heterogeneous pattern. Pharmacogenetics is the study of inherited differences in interindividual drug disposition and effects, with the goal of selecting the optimal drug therapy and dosage for each patient. Pharmacogenetics for cancer treatment is very significant, as cancer therapies exhibit severe systemic toxicity and unpredictable efficacy. There is presence of genetic polymorphisms in the genes which code for the metabolic enzymes and cellular targets for the majority of chemotherapy agents, but to predict the outcome of chemotherapy in patients is not currently possible for most treatments. A greater understanding of the genetic determinants of drug response can revolutionize the use of many medications. By identifying the patients at risk for severe toxicity, or those likely to benefit from a particular treatment, individualized cancer therapy can be achieved for most cancer patients. The prediction of cancer treatment outcome based on gene polymorphisms is becoming possible for many classes of chemotherapy agents, and the most clinically significant examples of chemotherapy agents are discussed in the chapter. However, further studies are needed in well characterized and larger cancer populations with proper validation of pharmacogenetic markers in experimental settings before application in clinical routine diagnostics.

Keywords: cancer, pharmacogenetics, polymorphism, chemotherapy, genetic variations

1. Introduction

The treatment of cancer has witnessed major advances which have resulted from the recent revolution in medical interventions. It is commonly observed in clinical settings that the same doses of medication cause considerable variations in efficacy and toxicity across human populations [1, 2]. These variations can lead to unpredictable life-threatening or even lethal adverse effects in cases receiving the medications [3, 4]. Genetic factors are important determinants for drug efficacy and toxicity since the interindividual variability in drug response cannot be explained only by physiological, life style, age, comedication, etc. factors (**Figure 1**). Pharmacogenetics is the study of how genetic inheritance influences response to drugs. The term "pharmacogenetics" was coined in the 1950s, with the discovery that there is an inherited basis for differences in the disposition and effects of drugs and xenobiotics [5]. The studies found that antimalarial drugs and certain foods (soy beans) cause hemolytic reactions in patients with glucose-6-phosphate dehydrogenase (G6PD) deficiency. The term pharmacogenomics and pharmacogenetics are often used interchangeably. Pharmacogenetics was first used in the literature in 1997 and ever since the developments in this field have been greatly facilitated by rapid progress in molecular technology, in particular, high throughput DNA sequencing, microarrays and genotyping [6].

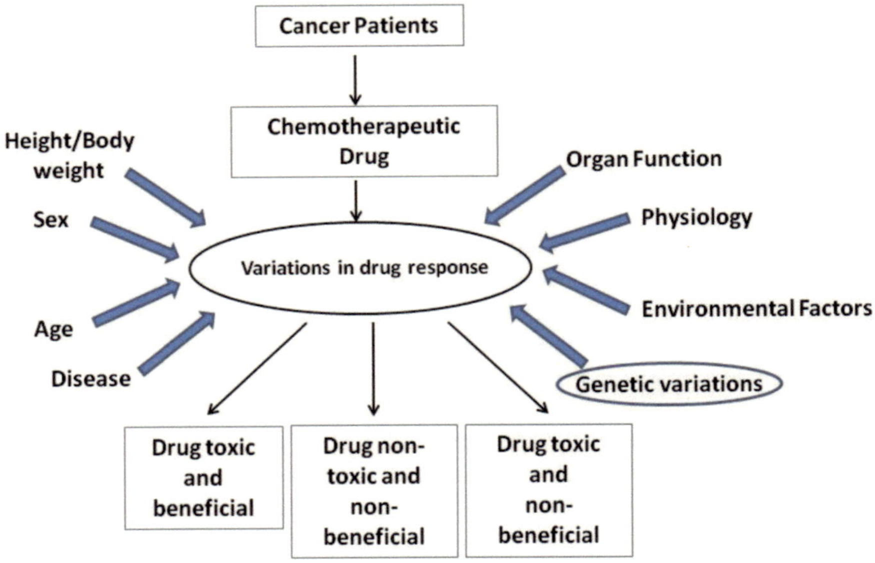

Figure 1.
Inter-individual variations in drug response.

Gene	Significant polymorphisms	Target drug	Action	Clinical relevance
TPMT	TPMT*2, *3A,*3B,*3C	6-Mercaptopurine (6-MP)	Increased levels of 6-MP	Myelotoxicity
DPD	DPD*2A	5-FU	Increased levels of 5-FU	Neurologic, hematological toxicities
UGT1A1	UGT1A1*28	Irinotecan	Increased levels of SN-38	Severe diarrhea, neutropenia
GST	Deletion, Ile105Val	Platinum agents	Increased DNA damage	Drug toxicity increased
XRCC1	Arg194Trp, Arg280His, Arg399Gln	Platinum agents	Increased DNA damage	Drug toxicity increased
ERCC1	K751Q	Platinum agents	Increased DNA damage	Drug toxicity increased

TPMT: thiopurine S-methyltransferase, DPD: dihydropyrimidine dehydrogenase, UGT1A1: UDP glucuronosyltransferase family 1 member A1, GST: glutathione S-transferase, XRCC1: X-ray repair cross complementing 1, ERCC1: excision repair 1, endonuclease non-catalytic subunit.

Table 1.
Pharmacogenetic biomarkers and their clinical impact.

Cancer	Biomarkers	Drugs
Breast	HER2	Trastuzumab, lapatinib
	ESR1	Exemestane, letrozole
Colorectal	KRAS	Cetuximab, panitumumab
	EGFR	Cetuximab, panitumumab
	DPD	5-Fluorouracil, capecitabine
	UGT1A1	Irinotecan
Lung	ALK	Crizotinib, ceritinib
	EGFR	Erlotinib, gefitinib
Gastrointestinal stromal tumor	c-Kit	Imatinib
Melanoma	BRAF	Vemurafenib, dabrafenib, trametinib
Pancreatic	EGFR	Erlotinib
Head and neck	EGFR	Cetuximab
Acute promyelocytic leukemia	PML-RARα	Arsenic trioxide, tretinoin
Cutaneous T-cell lymphoma	CD-25/IL2RA	Denileukin diftitox

HER2: human epidermal growth factor receptor 2, ESR1: estrogen receptor 1, KRAS: Ki-ras2 Kirsten rat sarcoma viral oncogene homolog, EGFR: epidermal growth factor receptor, DPD: dihydropyrimidine dehydrogenase, UGT1A1: UDP glucuronosyltransferase family 1 member A1, ALK: anaplastic lymphoma receptor tyrosine kinase, c-Kit: stem cell growth factor receptor, BRAF: B-Raf proto-oncogene, PML-RARα: promyelocytic leukemia/retinoic acid receptor alpha, CD-25/IL2RA: cluster of differentiation 25/interleukin 2 receptor subunit alpha.

Table 2.
Cancer pharmacogenetic biomarkers in FDA drug labelling.

Cancer pharmacogenetics has started getting a lot of attention due to the potential for individualisation of cancer therapy, minimizing toxicity, while maximizing efficacy. Cancer pharmacogenetics allows identification of patients at risk for severe toxicity, or those likely to benefit from a particular treatment and thus helps us move toward the ultimate goal of individualized cancer therapy. There are significant differences between cancer and other disease pharmacogenomics. In cancer, both germline genome of the patient and the somatic genome of the tumor are involved. The former is responsible for the inter-individual inherited genetic differences while the latter is due to accumulation of acquired somatic mutations resulting in inconsistent responses. Cancer pharmacogenomics also faces the problem of conducting human studies, availability of healthy volunteers for receiving cancer drugs and multigenic control of drug response. **Table 1** lists some of the major biomarkers which are associated with cancer treatment toxicities while **Table 2** lists some of the biomarkers associated with cancer treatments mentioned in US FDA-approved drug labels.

2. Candidate genes

The prediction of cancer treatment outcome based on gene polymorphisms is becoming a reality for many classes of chemotherapy agents, and the most clinically significant examples are discussed below.

2.1 Thiopurines

Thiopurines are a family of drugs that includes 6-mercaptopurine (6-MTP) which is a daily component of maintenance therapy for childhood acute lymphoblastic leukemia treatment [7], thioguanine and azathioprine. There are three major metabolic pathways for 6-MTP namely activation of 6-MTP into 6-TGN (activating and cytotoxic route for thiopurines) by hypoxanthine guanine phosphoribosyl transferase (HGPRT), inactivation of 6-MTP into thiouric acid via oxidation catalyzed by xanthine oxidase and inactivation of 6-MTP via S-methylation of the thiol moiety in the liver and red blood cells by thiopurine methyltransferase (TPMT). This methylation shunts the active drug away from TGN formation. Thiopurines are inactive prodrugs that require metabolism to thioguanine nucleotides (TGN) to exert cytotoxicity by incorporation of TGN into DNA. This activation is catalyzed by methylation by thiopurine methyltransferase (TPMT) of the thiopurine agents azathioprine, mercaptopurine, and thioguanine [7, 8], thereby shunting drug away from TGN formation. Genetic variants present in TPMT may alter the treatment response in cases receiving chemotherapeutic drugs. TPMT exhibits huge variations in enzyme activity with a major portion of population (90%) exhibiting high activity while about 10% have intermediate activity, and 0.3% have low or no detectable enzyme activity [9, 10]. Out of several genetic variants, TPMT*2 (238G>C), *3A (460G>A and 719A>G), *3B (460G>A), and *3C (719A>G) account for about 95% of intermediate or low enzyme activity cases [7, 11–14]. Caucasians have more prevalence of TPMT polymorphisms and a trimodal distribution of TPMT enzyme activity while southeast Asians have less prevalence and a unimodal distribution [15–17].

Several studies have shown that TPMT-deficient patients are at very high risk of developing severe hematopoietic toxicity if treated with conventional doses of thiopurines [18, 19]. Studies have also been carried out to show that patients who are heterozygous at the TPMT locus are at intermediate risk of dose-limiting toxicity [20, 21]. In one of our studies, a poor treatment response was observed in head and neck cancer patients receiving chemotherapy with cisplatin and 5-FU which might be due to the higher intracellular concentration of cisplatin due to lower or intermediate TPMT enzyme activity [22]. Liu et al. [23] examined primary erythrocyte TPMT activity in children with leukemia in a genome-wide association study and found that TPMT was the only gene that reached genome-wide significance. In another study of 67 patients treated with azathioprine for rheumatic disease, 6 patients (9%) were heterozygous for mutant TPMT alleles, and therapy was discontinued in 5 of 6 patients because of low leukocyte count within 1 month of starting treatment [20].

2.2 5-Fluorouracil (5-FU)

5-Fluorouracil (5-FU) is a uracil analog that is widely used to treat solid tumors, such as colorectal and breast cancer and requires activation to 5-fluoro-2-deoxyuridine monophosphate (5-FdUMP). At least 85% of 5-FU is inactivated by dihydropyrimidine dehydrogenase (DPD) to dihydrofluorouracil in the liver. [24] 5-FdUMP acts by inhibiting the tumor cell replication via inhibition of thymidylate synthase (TS), an enzyme that is required for de novo pyrimidine synthesis. DPD inactivates 5-FU in the liver and has huge differences in activity among individuals leading to excessive amounts of 5-FdUMP in patients with low activity which causes gastrointestinal, hematopoietic, and neurological toxicities [25–32].

The DPD gene has several reported polymorphisms associated with reduced DPD activity [32, 33]. In general, 3–5% of individuals are heterozygous carriers of mutations that inactivate DPD, and 0.1% of individuals are homozygous for mutations that inactivate DPD [27, 34–36]. DPD*2A allele is caused by a G>A transition at a GT splice donor site flanking exon 14 of the DPD gene (IVS14+1G>A). A decreased DPD activity has been found to be associated with severe or fatal toxicity from standard doses of 5-FU [37]. Another mutation at codon 534 leads to a 1601G>A nucleotide change. In one of our study, head and neck cancer patients exhibited a poor treatment response which had IVS14+1G>A genetic variant. [22] Similarly, it was also reported that IVS14+1G>A was associated with increased toxicity and poor treatment response in patients of invasive ductal carcinoma and head and neck cancer. Zhao et al. [38] found that DPD variant c.85T>C (rs1801265, DPYD*9A) was associated with treatment outcome in acute lymphoblastic leukemia.

2.3 Irinotecan

Irinotecan, a topoisomerase I inhibitor, is used to treat various solid tumors, and requires activation by carboxylesterase to its active metabolite, SN-38. The toxicities associated with Irinotecan, namely diarrhea and leucopenia, are due to increased levels of SN-38. UDP-glucuronosyltransferase 1A1 (UGT1A1) present in liver metabolizes SN-38 by glucuronidation to produce the more polar and inactive SN-38 glucuronide, which is removed in the bile and urine [39]. In chemotherapy, a decreased rate of glucuronidation has been shown to be an important factor in prediction of toxicity. The rate of glucuronidation is reduced as a consequence of reduced transcription rate due to abnormal dinucleotide repeat sequences (5–8 repeats) within the TATA box of the UGT1A1 gene promoter [40]. An inverse relationship exists between the number of TA repeats and the UGT1A1 transcription rate. The variant allele UGT1A1*28 results from the presence of seven repeats, instead of the wild-type number of six. The UGT1A1*28 allele is associated with reduced UGT1A1 expression, and leads to reduced SN-38 glucuronidation [41].

(TA)n TAA promoter polymorphisms are more frequent in Caucasians than in Asians which have more missense polymorphisms in the exons [42]. Studies have shown that the UGT1A1*28 allele leads to significantly increased amounts of the active metabolite SN-38, and consequently an increased chance of developing side effects such as diarrhea and leukopenia during irinotecan therapy. [41, 43]. In one study of 20 patients with solid tumors treated by irinotecan, severe toxicity was observed in UGT1A1*28 heterozygotes and homozygotes [41]. In another retrospective study of 118 cancer patients treated with irinotecan, a significant proportion of the 26 patients suffered from severe diarrhea or neutropenia. Upon examination, all 26 were UGT1A1*28 homozygotes or heterozygotes (15 and 31%, respectively), whereas only 3% UGT1A1*28 homozygotes and 11% UGT1A1*28 homozygotes were found among 92 patients without toxicity [43]. Font et al. [44] reported that 34% of non-small cell lung carcinoma (34%) patients with the common genotype achieved disease control (partial response or stable disease) compared with 13 of 24 patients (54%) with the variant genotypes.

2.4 Platinum agents

Platinum agents like cisplatin, carboplatin and oxaliplatin act by inhibiting cell replication as a result of formation of DNA adducts. However, sometimes the effect of platinum agents is compromised as a result of decreased drug accumulation, detoxification, reduced or no DNA adduct formation and an increased activity of DNA repair system. One of the factors that can influence response to platinum chemotherapy agents is polymorphisms in glutathione (GSH)-dependent enzymes. Glutathione-S-transferases (GSTs) catalyze the conjugation of GSH to platinum agents, forming less toxic and more water-soluble conjugates. There are five subclasses of the GST family (GSTA1, GSTP1, GSTM1, GSTT1, and GSTZ1) [61] that influence cytotoxicity to a variety of chemotherapeutic agents [45].

Several genetic variants exist in the GSTs which may lead to complete absence (GSTM1 and GSTT1) or partially deficient enzyme (GSTP1) activity. Ethnic differences are reported in the distribution of null or variant allele frequencies of GSTM1, GSTT1 and GSTP1. Studies from our laboratory have shown association of polymorphism in drug metabolizing cytochrome P450s (CYPs) and GSTs with head and neck cancer [46]. Studies also revealed significant increase in head and neck cancer risk in cases with null genotypes of GSTM1 or GSTT1, though inconsistent reports are also available. Likewise, no consistent data is available on the association of GSTP1 polymorphism with head and neck cancer risk [47]. Further, site specificity is also reported in the expression of GSTs in the squamous mucosa of head and neck which may lead to the differences in the susceptibility when analyzed according to the tumor location. An association between the GSTM1 and GSTT1 null genotype for non-laryngeal upper aero-digestive tract (UADT) or oral cancer risk was reported in smokers or tobacco chewers. In contrast, no site specific differences in the distribution of GST variant forms have also been observed in few studies [48].

The null genotypes for GSTM1 or GSTT1 were associated with a reduction in risk of relapse in several tumor types treated with chemotherapy such as acute lymphoblastic leukemia, acute myeloblastic leukemia, breast cancer, ovarian cancer, and lung cancer. In addition to null phenotypes, single nucleotide polymorphisms (SNPs) also affect response to chemotherapy and survival of patients as seen in breast cancer patients with an I105V SNP in the GSTP1 gene. Women with the low-activity VV genotype had better survival upon cyclophosphamide-based chemotherapy [49]. Dasgupta et al. [50] compared the role of the I105V genotype in multiple myeloma treated with standard or high dose chemotherapy and found that the patients with the 105VV homozygote allele had an improved progression free survival. The substitution of isoleucine with valine at position 105 reduces enzyme activity against alkylating agents. Stoehlmacher et al. [51] showed that SNP in GSTP1 was associated with overall survival in 107 patients with metastatic colorectal cancer who received 5-FU/oxaliplatin combination chemotherapy. In this study, 10 patients (9%) were homozygous for valine, 45 patients (42%) were heterozygous, and 52 (49%) were homozygous for isoleucine. Interestingly, GSTM1 and GSTT1 mutations that abolish enzyme activity had no predictive power for patient outcome.

Anticancer agents act by causing DNA damage in tumor cells which is subsequently repaired by the DNA repair machinery of the cell. Thus, more the active DNA repair system, less will be the treatment outcome. XRCC1 is a prominent gene involved in DNA repair via the base excision repair pathway which repairs single strand breaks through interaction of XRCC1 with PARP-1, PNK, Polb, and Lig3a [52]. XRCC1 has several genetic variants out of which the prominent ones are Arg194Trp on exon 6, Arg280His on exon 9, and Arg399Gln on exon 10 [53]. A study conducted by Quintela-Fandino et al. [54] in head and neck cancer cases found that XRCC1 Gln/Gln was responsible for 61.5% of cases with complete response. The role of XRCC1-Gln399Gln genotype was also investigated by Duell et al. [55]. It was reported that the allele results in high rate of sister chromatid exchange after exposure to ionizing radiation in human lymphocytes. There are also reports which suggest the role of XRCC1 G28152A Arg399Gln polymorphism in development of lower grade of fibrosis as a result of radiotherapy in 60 nasopharyngeal cancer patients [56]. Mahimkar et al. [57] studied clinical outcome in advanced oral cancer patients treated with postoperative radiotherapy and did not observe a significant association between polymorphisms of XRCC1 and clinical outcome. Zhai et al. [58] observed that Codon399 Gln/Gln allele was associated with a higher tumor regression ratio after radiotherapy for primary nasopharyngeal neoplasm and metastatic lymph nodes. Ghazali et al. [59] conducted a systematic review and found that risk of severe acute mucositis was associated with the G allele of XRCC1 (1196A>G) in head and neck cancer patients treated with radiotherapy alone or chemotherapy.

Excision-repair cross-complementing 1 (ERCC1) gene encodes a helicase which is required for the nucleotide excision repair pathway. Several polymorphisms in ERCC1 which result in differing DNA repair capacities have been identified. Lowered mRNA production was observed as a result of a silent C118T SNP in ovarian carcinoma cell lines [60]. The TT genotype resulted in a reduction in codon usage by half with a reduction in ERCC1 mRNA production and therefore be associated with reduced DNA repair capacity [61]. Platinum is a standard chemotherapy for advanced non-small cell lung cancer (NSCLC), and platinum-induced DNA lesions are repaired by ERCC1. Studies have shown that patients homozygous for the ERCC1 118C allele demonstrated a significantly better survival. In colorectal carcinoma patients treated with 5-fluorouracil and oxaliplatin, the K751Q SNP of the ERCC2 (Xeroderma pigmentosum group D gene, XPD) determined in peripheral blood lymphocytes was of prognostic relevance. The patients having KK homozygotes responded more frequently to chemotherapy and lived significantly longer than did heterozygotes or QQ homozygotes [62]. Time to progression was significantly higher in cisplatinum-treated patients with non-small cell lung cancer harboring the K751Q ERCC2 genotype than those harboring the K751K genotype. However, contradictory results on the association of ERCC2/XPD variant alleles with decreasing overall survival of non-small cell lung cancer patients after cisplatinum-based therapy were also reported [63]. A nonsynonymous SNP, altering a lysine to glutamine at codon 751 of the XPD protein, was significantly associated with treatment outcome in patients with metastatic colorectal cancer [62].

3. Pharmacogenetics: challenges and next generation approaches

The current pharmacogenetics approaches face many stumbling blocks. Candidate gene-based approaches do not provide a reliable prediction of tumor drug response and normal tissue toxicity because of a lack of understanding of the precise role of all participating factors. Genome wide association study provide a more robust platform for pharmacogenetic analysis as has been demonstrated by Watters et al. [64]. A number of other issues plague SNP genotyping in the clinical settings such as quality control which is due phenotypic heterogeneity, a long duration involved in validation of pharmacogenetic markers in experimental settings, the combined effects of many low-risk polymorphisms, selection of the most appropriate panel of SNPs, analyzing the correlation between genotype, gene expression, and enzyme activity, criteria for risk assessment and thresholds, consideration of ethnic variations as the distribution and frequency of SNPs vary among different ethnic groups which makes it difficult to extrapolate the findings of one group on another [65]. Newer targeted therapies are also gaining popularity. Trastuzumab (herceptin), a humanized recombinant monoclonal antibody (IgG) targets Human Epidermal Growth Factor Receptor 2 (HER2), Gefitinib (Iressa) inhibits the tyrosine kinase activity of the Epidermal Growth Factor Receptor, Bevacizumab (Avastin) is an anti-angiogenesis agent, the addition of which to standard chemotherapy regimens has shown improved response rates and survival rates in the treatment of metastatic colorectal cancer [66]. Likewise, Cetuximab (Erbitux), a monoclonal antibody, targeting EGFR has also shown promising results in colorectal cancer and head and neck cancers.

Future developments in some key areas will play a critical role in deciding the overall influence of pharmacogenetics data on therapeutic decisions. Improvements are needed in genome-wide technologies such as development of gene expression arrays, high throughput technologies, SNP chips, genome-wide scans which could potentially identify previously unidentified, functionally important candidate genes and SNPs. Mouse models could be used for genome-wide scans in offspring from phenotypically distinct mice from resistant and susceptible strains. Knockout and transgenic techniques could also be used for establishing the key elements that contribute to drug response and disposition. Candidate gene approach could be enhanced by knowledge gained from genome wide techniques and by incorporating a metabolic pathway approach. The cost of SNP/genomic technology should reduce which needs to be counterbalanced by the huge costs incurred due to adverse drug reactions/toxicities. For inclusion of a genetic test into clinical practice, it must provide reliable, predictive, and actionable information that would have otherwise been unknown [67]. Before clinical implementation, strong evidence from randomized controlled clinical trials is needed.

The future of pharmacogenetics should focus on specimen collection of both germline and tumor DNA from early and later phase clinical trials with prospectively collected efficacy and toxicity data which will be vital in the discovery and validation of pharmacogenomic associations. At next steps, genes that have undergone replication and validation should be assessed for clinical implementation. A large retrospective case-control validation and replication studies and Phase II biomarker-driven clinical trials may allow for a more efficient and rapid method of translation from bench to bedside.

4. Conclusions

The major problems of cancer chemotherapy are the development of drug resistance and the severe side effects. Since many chemotherapeutic agents have modest tumor specificity, normal tissues are also damaged. This prevents the application of sufficient high doses of drugs to eradicate the less sensitive tumor cell populations. Thereby, tumors develop drug resistance that leads to treatment failure and fatal consequences for patients. Genetic variations in genes have explained a great deal of interindividual variation in response and toxicity of anticancer drugs. Cancer treatment utilizes multiple therapeutic agents with a wide variety of toxicities, often with narrow therapeutic indices. Pharmacogenetics has the potential to revolutionize cancer therapy. Though there has been substantial success in situations where single genes play a large role in overall drug response, the future of cancer treatment lies in whole-genome approaches. Reduction of the toxicogenetic and toxicogenomic side effects has been one of the major goals in the search for new anticancer drugs and therapy protocols. SNP genotyping should be introduced into clinical settings to facilitate clinical decision making regarding treatment strategies to avoid adverse drug reactions while achieving the best drug response. Few of the studies discussed do provide a stronger scientific basis for the use of genomic information for the individualization of cancer therapy based on a patient's genetic profile.

Conflict of interest

The author declares no conflict of interest.

DOI: http://dx.doi.org/10.5772/intechopen.83424

Author details

Munindra Ruwali
Amity Institute of Biotechnology, Amity University Haryana, Gurgaon, India

*Address all correspondence to: munindraruwali@gmail.com

References

[1] Evans WE, Relling MV. Pharmacogenomics: Translating functional genomics into rational therapeutics. Science. 1999;**286**:487-491

[2] Fagerlund TH, Braaten O. No pain relief from codeine? An introduction to pharmacogenomics. Acta Anaesthesiologica Scandinavica. 2001;**45**:140-149

[3] Rothenberg ML, Meropol NJ, Poplin EA, van Cutsem E, Wadler S. Mortality associated with irinotecan plus bolus fluorouracil/leucovorin: Summary findings of an independent panel. Journal of Clinical Oncology. 2001;**19**:3801-3807

[4] Sargent DJ, Niedzwiecki D, O'Connell MJ, Schilsky RL. Recommendation for caution with irinotecan, fluorouracil, and leucovorin for colorectal cancer. The New England Journal of Medicine. 2001;**345**:144-145

[5] Vogel F. Moderne probleme der humangenetik. Ergebnisse der Inneren Medizin und Kinderheilkunde. 1959;**12**:52-125

[6] Weinshilboum R, Wang L. Pharmacogenomics: Bench to bedside. Nature Reviews. Drug Discovery. 2004;**3**:739-748

[7] McLeod HL, Krynetski EY, Relling MV, Evans WE. Genetic polymorphism of thiopurine methyltransferase and its clinical relevance for childhood acute lymphoblastic leukemia. Leukemia. 2000;**14**:567-572

[8] Krynetski EY, Evans WE. Pharmacogenetics of cancer therapy: Getting personal. American Journal of Human Genetics. 1998;**63**:11-16

[9] Weinshilboum RM, Sladek SL. Mercaptopurine pharmacogenetics: Monogenic inheritance of erythrocyte thiopurine methyltransferase activity. American Journal of Human Genetics. 1980;**32**:651-662

[10] McLeod HL, Relling MV, Liu Q, Pui CH, Evans WE. Polymorphic thiopurine methyltransferase in erythrocytes is indicative of activity in leukemic blasts from children with acute lymphoblastic leukemia. Blood. 1995;**85**:1897-1902

[11] Krynetski EY, Schuetz JD, Galpin AJ, Pui CH, Relling MV, Evans WE. A single point mutation leading to loss of catalytic activity in human thiopurine S-methyltransferase. Proceedings of the National Academy of Sciences of the United States of America. 1995;**92**:949-953

[12] Tai HL, Krynetski EY, Yates CR, Loennechen T, Fessing MY, Krynetskaia NF, et al. Thiopurine S-methyltransferase deficiency: Two nucleotide transitions define the most prevalent mutant allele associated with loss of catalytic activity in Caucasians. American Journal of Human Genetics. 1996;**58**:694-702

[13] Otterness D, Szumlanski C, Lennard L, Klemetsdal B, Aarbakke J, Park-Hah JO, et al. Human thiopurine methyltransferase pharmacogenetics: Gene sequence polymorphisms. Clinical Pharmacology and Therapeutics. 1997;**62**:60-73

[14] Yates CR, Krynetski EY, Loennechen T, Fessing MY, Tai HL, Pui CH, et al. Molecular diagnosis of thiopurine S-methyltransferase deficiency: Genetic basis for azathioprine and mercaptopurine intolerance. Annals of Internal Medicine. 1997;**126**:608-614

[15] Lennard L, Lilleyman JS, van Loon J, Weinshilboum RM. Genetic variation in response to 6-mercaptopurine for childhood acute lymphoblastic leukaemia. Lancet. 1990;**336**:225-229

[16] Weinshilboum R. Thiopurine pharmacogenetics: Clinical and molecular studies of thiopurine methyltransferase. Drug Metabolism and Disposition. 2001;**29**:601-605

[17] Sahasranaman S, Howard D, Roy S. Clinical pharmacology and pharmacogenetics of thiopurines. European Journal of Clinical Pharmacology. 2008;**64**:753-767

[18] Lennard L, VanLoon JA, Weinshilboum RM. Pharmacogenetics of acute azathioprine toxicity: Relationship to thiopurine methyltransferase genetic polymorphism. Clinical Pharmacology and Therapeutics. 1989;**46**:149-154

[19] Evans WE, Horner M, Chu YQ, Kalwinsky D, Roberts WM. Altered mercaptopurine metabolism, toxic effects, and dosage requirement in a thiopurine methyltransferase-deficient child with acute lymphocytic leukemia. The Journal of Pediatrics. 1991;**119**:985-989

[20] Black AJ, McLeod HL, Capell HA, Powrie RH, Matowe LK, Pritchard SC, et al. Thiopurine methyltransferase genotype predicts therapy-limiting severe toxicity from azathioprine. Annals of Internal Medicine. 1998;**129**:716-718

[21] Relling MV, Hancock ML, Rivera GK, Sandlund JT, Ribeiro RC, Krynetski EY, et al. Mercaptopurine therapy intolerance and heterozygosity at the thiopurine S-methyltransferase gene locus. Journal of the National Cancer Institute. 1999;**91**:2001-2008

[22] Dhawan A, Ruwali M, Pant MC, Rahman Q, Parmar D. Association of genetic variability in enzymes metabolizing chemotherapeutic agents with treatment response in head and neck cancer cases. Asia-Pacific Journal of Clinical Oncology. 2017;**13**:e11-e20

[23] Liu C, Yang W, Pei D, Cheng C, Smith C, Landier W, et al. Genomewide approach validates thiopurine methyltransferase activity is a monogenic pharmacogenomic trait. Clinical Pharmacology and Therapeutics. 2017;**101**:373-381

[24] Heggie GD, Sommadossi J-P, Cross DS, Huster WJ, Diasio RB. Clinical pharmacokinetics of 5 fluorouracil and its metabolism in plasma, urine and bile. Cancer Research. 1987;**47**:2203-2306

[25] Etienne MC, Lagrange JL, Dassonville O, Fleming R, Thyss A, Renee N, et al. Population study of dihydropyrimidine dehydrogenase in cancer patients. Journal of Clinical Oncology. 1994;**12**:2248-2253

[26] Lu Z, Zhang R, Diasio RB. Dihydropyrimidine dehydrogenase activity in human peripheral blood mononuclear cells and liver: Population characteristics, newly identified deficient patients, and clinical implication in 5-fluorouracil chemotherapy. Cancer Research. 1993;**53**:5433-5438

[27] Johnson MR, Hageboutros A, Wang K, High L, Smith JB, Diasio RB. Life-threatening toxicity in a dihydropyrimidine dehydrogenase-deficient patient after treatment with topical 5-fluorouracil. Clinical Cancer Research. 1999;**5**:2006-2011

[28] Diasio RB, Beavers TL, Carpenter JT. Familial deficiency of dihydropyrimidine dehydrogenase. Biochemical basis for familial pyrimidinemia and severe 5-fluorouracil-induced toxicity. The Journal of Clinical Investigation. 1988;**81**:47-51

[29] Diasio RB. Clinical implications of dihydropyrimidine dehydrogenase on 5-FU pharmacology. Oncology (Huntingt.). 2001;**15**:21-26 (discussion 27)

[30] van Kuilenburg AB, Muller EW, Haasjes J, Meinsma R, Zoetekouw L, Waterham HR, et al. Lethal outcome of a patient with a complete dihydropyrimidine dehydrogenase (DPD) deficiency after administration of 5-fluorouracil: Frequency of the common IVS14+1G>A mutation causing DPD deficiency. Clinical Cancer Research. 2001;7:1149-1153

[31] Innocenti F, Ratain MJ. Update on pharmacogenetics in cancer chemotherapy. European Journal of Cancer. 2002;**38**:639-644

[32] Wei X, McLeod HL, McMurrough J, Gonzalez FJ, Fernandez-Salguero P. Molecular basis of the human dihydropyrimidine dehydrogenase deficiency and 5-fluorouracil toxicity. The Journal of Clinical Investigation. 1996;**98**:610-615

[33] McLeod HL, Collie-Duguid ES, Vreken P, Johnson MR, Wei X, Sapone A, et al. Nomenclature for human DPYD alleles. Pharmacogenetics. 1998;**8**:455-459

[34] Gonzalez FJ, Fernandez-Salguero P. Diagnostic analysis, clinical importance and molecular basis of dihydropyrimidine dehydrogenase deficiency. Trends in Pharmacological Sciences. 1995;**16**:325-327

[35] Ridge SA, Sludden J, Brown O, Robertson L, Wei X, Sapone A, et al. Dihydropyrimidine dehydrogenase pharmacogenetics in Caucasian subjects. British Journal of Clinical Pharmacology. 1998;**46**:151-156

[36] Ridge SA, Sludden J, Wei X, Sapone A, Brown O, Hardy S, et al. Dihydropyrimidine dehydrogenase pharmacogenetics in patients with colorectal cancer. British Journal of Cancer. 1998;**77**:497-500

[37] Van Kuilenburg AB, Meinsma R, Zoetekouw L, Van Gennip AH. High prevalence of the IVS14+1G>A mutation in the dihydropyrimidine dehydrogenase gene of patients with severe 5-fluorouracil associated toxicity. Pharmacogenetics. 2002;**12**:555-558

[38] Zhao XQ, Cao WJ, Yang HP, Yang XW, Tang P, Sun L, et al. DPYD gene polymorphisms are associated with risk and chemotherapy prognosis in pediatric patients with acute lymphoblastic leukemia. Tumour Biology. 2016;**37**:10393-10402

[39] Gupta E, Lestingi TM, Mick R, Ramirez J, Vokes EE, Ratain MJ. Metabolic fate of irinotecan in humans: Correlation of glucuronidation with diarrhea. Cancer Research. 1994;**54**:3723-3725

[40] Innocenti F, Iyer L, Ratain MJ. Pharmacogenetics of anticancer agents: Lessons from amonafide and irinotecan. Drug Metabolism and Disposition. 2001;**29**:596-600

[41] Iyer L, Das S, Janisch L, Wen M, Ramirez J, Karrison T, et al. UGT1A1*28 polymorphism as a determinant of irinotecan disposition and toxicity. The Pharmacogenomics Journal. 2002;**2**:43-47

[42] Beutler E, Gelbart T, Demina A. Racial variability in the UDP-glucuronosyltransferase 1 (UGT1A1) promoter: A balanced polymorphism for regulation of bilirubin metabolism? Proceedings of the National Academy of Sciences of the United States of America. 1998;**95**:8170-8174

[43] Ando Y, Saka H, Ando M, Sawa T, Muro K, Ueoka H, et al. Polymorphisms of UDP-glucuronosyltransferase gene and irinotecan toxicity: A pharmacogenetic analysis. Cancer Research. 2000;**60**:6921-6926

[44] Font A, Sánchez JM, Tarón M, Martinez-Balibrea E, Sánchez JJ, Manzano JL, et al. Weekly regimen

of irinotecan/docetaxel in previously treated non-small cell lung cancer patients and correlation with uridine diphosphate glucuronosyltransferase 1A1 (UGT1A1) polymorphism. Investigational New Drugs. 2003;**21**:435-443

[45] Chao CC, Huang YT, Ma CM, Chou WY, Lin-Chao S. Overexpression of glutathione S-transferase and elevation of thiol pools in a multidrug-resistant human colon cancer cell line. Molecular Pharmacology. 1992;**41**:69-75

[46] Ruwali M, Parmar D. Association of functionally important polymorphisms in cytochrome P450s with squamous cell carcinoma of head and neck. Indian Journal of Experimental Biology. 2010;**48**:651-665

[47] Singh M, Shah PP, Singh AP, Ruwali M, Mathur N, Pant MC, et al. Association of genetic polymorphisms in glutathione S-transferases and susceptibility to head and neck cancer. Mutation Research. 2008;**638**:184-194

[48] Peters ES, McClean MD, Marsit CJ, Luckett B, Kelsey KT. Glutathione S-transferase polymorphisms and the synergy of alcohol and tobacco in oral, pharyngeal, and laryngeal carcinoma. Cancer Epidemiology, Biomarkers & Prevention. 2006;**15**:2196-2202

[49] Sweeney C, McClure GY, Fares MY, Stone A, Coles BF, Thompson PA, et al. Association between survival after treatment for breast cancer and glutathione S-transferase P1 Ile105Val polymorphism. Cancer Research. 2000;**60**:5621-5624

[50] Dasgupta RK, Adamson PJ, Davies FE, Rollinson S, Roddam PL, Ashcroft AJ, et al. Polymorphic variation in GSTP1 modulates outcome following therapy for multiple myeloma. Blood. 2003;**102**:2345-2350

[51] Stoehlmacher J, Park DJ, Zhang W, Groshen S, Tsao-Wei DD, Yu MC, et al. Association between glutathione S-transferase P1, T1, and M1 genetic polymorphism and survival of patients with metastatic colorectal cancer. Journal of the National Cancer Institute. 2002;**94**:936-942

[52] Ramachandran S, Ramadas K, Hariharan R, Rejnish Kumar R, Radhakrishna Pillai M. Single nucleotide polymorphisms of DNA repair genes XRCC1 and XPD and its molecular mapping in Indian oral cancer. Oral Oncology. 2006;**42**:350-362

[53] Bao Y, Jiang L, Zhou JY, Zou JJ, Zheng JY, Chen XF, et al. XRCC1 gene polymorphisms and the risk of differentiated thyroid carcinoma (DTC): A meta-analysis of case-control studies. PLoS One. 2013;**8**:e64851

[54] Quintela-Fandino M, Hitt R, Medina PP, Gamarra S, Manso L, Cortes-Funes H, et al. DNA-repair gene polymorphisms predict favorable clinical outcome among patients with advanced squamous cell carcinoma of the head and neck treated with cisplatin based induction chemotherapy. Journal of Clinical Oncology. 2006;**24**:4333-4339

[55] Duell EJ, Wiencke JK, Cheng TJ, Varkonyi A, Zuo ZF, Ashok TD, et al. Polymorphisms in the DNA repair genes XRCC1 and ERCC2 and biomarkers of DNA damage in human blood mononuclear cells. Carcinogenesis. 2000;**21**:965-971

[56] Alsbeih G, Al-Harbi N, Al-Hadyan K, El-Sebaie M, Al-Rajhi N. Association between normal tissue complications after radiotherapy and polymorphic variations in TGFB1 and XRCC1 genes. Radiation Research. 2010;**173**:505-511

[57] Mahimkar MB, Samant TA, Kannan S, Tulsulkar J, Pai PS, Anantharaman D. Polymorphisms in

GSTM1 and XPD genes predict clinical outcome in advanced oral cancer patients treated with postoperative radiotherapy. Molecular Carcinogenesis. 2012;**51**:E94-E103

[58] Zhai XM, Hu QC, Gu K, Wang JP, Zhang JN, Wu YW. Significance of XRCC1 Codon399 polymorphisms in Chinese patients with locally advanced nasopharyngeal carcinoma treated with radiation therapy. Asia-Pacific Journal of Clinical Oncology. 2016;**12**:e125-e132

[59] Ghazali N, Shaw RJ, Rogers SN, Risk JM. Genomic determinants of normal tissue toxicity after radiotherapy for head and neck malignancy: A systematic review. Oral Oncology. 2012;**48**:1090-1100

[60] Yu JJ, Mu C, Lee KB, Okamoto A, Reed EL, Bostick-Bruton F, et al. A nucleotide polymorphism in ERCC1 in human ovarian cancer cell lines and tumor tissues. Mutation Research. 1997;**382**:13-20

[61] Yu JJ, Lee KB, Mu C, Li Q, Abernathy TV, Bostick-Bruton F, et al. Comparison of two human ovarian carcinoma cell lines (A2780/CP70 and MCAS) that are equally resistant to platinum, but differ at codon 118 of the ERCC1 gene. International Journal of Oncology. 2000;**16**:555-560

[62] Park DJ, Stoehlmacher J, Zhang W, Tsao-Wei DD, Groshen S, Lenz HJ. A Xeroderma pigmentosum group D gene polymorphism predicts clinical outcome to platinum-based chemotherapy in patients with advanced colorectal cancer. Cancer Research. 2001;**61**:8654-8658

[63] Ryu JS, Hong YC, Han HS, Lee JE, Kim S, Park YM, et al. Association between polymorphisms of ERCC1 and XPD and survival in non-small-cell lung cancer patients treated with cisplatin combination chemotherapy. Lung Cancer. 2004;**44**:311-316

[64] Watters JW, Kraja A, Meucci MA, Province MA, McLeod HL. Genome-wide discovery of loci influencing chemotherapy cytotoxicity. Proceedings of the National Academy of Sciences of the United States of America. 2004;**101**:11809-11814

[65] Oscarson M. Pharmacogenetics of drug metabolising enzymes: Importance for personalised medicine. Clinical Chemistry and Laboratory Medicine. 2003;**41**:573-580

[66] Hurwitz H, Fehrenbacher L, Novotny W, Cartwright T, Hainsworth J, Heim W, et al. Bevacizumab plus irinotecan, fluorouracil, and leucovorin for metastatic colorectal cancer. The New England Journal of Medicine. 2004;**350**:2335-2342

[67] McLeod HL. Cancer pharmacogenomics: Early promise, but concerted effort needed. Science. 2013;**339**:1563-1566

Chapter 4

CTCs as Liquid Biopsy: Where Are We Now?

Laure Cayrefourcq and Catherine Alix-Panabières

Abstract

A few years ago, the analysis of circulating tumor cells (CTCs) in the blood of patients with cancer was defined by the term "real-time liquid biopsy." Blood samples can be obtained and analyzed at the time of diagnosis and repeatedly during the systemic treatment. The analysis of the *liquid biopsy* has provided new insights into the biology of metastasis with important implications for the clinical management of cancer patients. In this review, we updated all technical strategies developed to improve enrichment, detection, and characterization of CTCs. We also focused on their biological properties as well as on their clinical relevance in different cancer types. At the end, we opened the discussion to all the other circulating biomarkers used as *liquid biopsy*.

Keywords: circulating tumor cells, *liquid biopsy*, clinical relevance, circulating biomarkers, precision medicine

1. Introduction

A few years ago, the analysis of circulating tumor cells (CTCs) in the blood of patients with cancer was defined by the term "*liquid biopsy*" [1]. Blood samples can be obtained and analyzed at the time of diagnosis and during the systemic treatment. Detection of CTCs in circulation gives important information on the molecular properties of tumor lesions. This information contributes to the early detection of metastatic lesions and participates in the personalized treatment of cancer patients such as prognostic evaluation, stratification of patients for targeted therapies, real-time monitoring of treatment efficacy, identification of therapeutic target, and resistance mechanism.

The analysis of the *liquid biopsy* has provided new insights into the biology of metastasis with important implications for the clinical management of cancer patients (**Figure 1**). Numerous clinical studies and meta-analyses including large cohorts of patients have shown that the number of CTCs is an important indicator of the risk of progression or death in patients with metastatic solid cancer (e.g., breast, prostate, colon, etc.) [2–6].

Despite the remarkable advances made in recent years, so far, *liquid biopsy* analyses are rarely implemented in routine patient testing. In-depth investigation of CTCs remains technically challenging. CTCs occur at the very low concentrations of one tumor cell in the background of millions of blood cells. Their identification and characterization require extremely sensitive and specific analytic methods. Moreover, up to now, results obtained with liquid biopsy analysis did not lead yet to

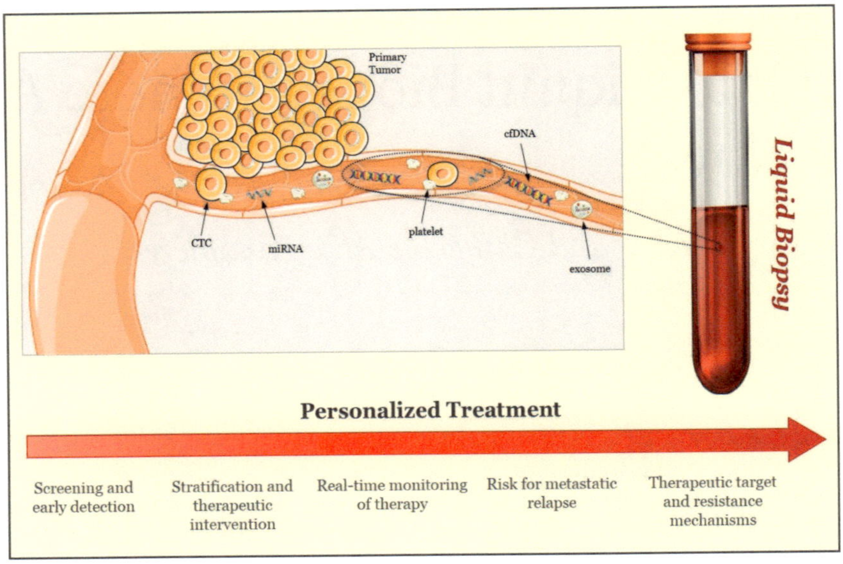

Figure 1.
From the blood sample toward the precision medicine in cancer patients.

validated guidelines for treatment and patient management. Nevertheless, technical advances and encouraging clinical studies demonstrated that *liquid biopsy* holds great promise for revolutionizing cancer diagnostics in a soon future.

Here, we will outline the advantages and challenges of CTCs as *liquid biopsy* in oncology by discussing the strategies for enrichment, detection, and characterization linked to the biology of these cells. Moreover, the potential of CTC analysis for clinical utility will be argued as well as other circulating biomarkers.

2. Technical strategies for enrichment, detection, and characterization of CTCs

At the moment, in-depth investigation of CTCs still remains technically challenging as they are every rare events in blood circulation. Their identification and characterization require extremely sensitive and specific analytic methods, which are usually a combination of enrichment and detection procedures (**Figure 2**). The different strategies to analyze CTCs are described in this chapter, and all the advantages/disadvantages plus the commercial status are summarized in **Table 1**.

2.1 Strategies for CTC enrichment

Up to date, a large panel of technologies was designed to enrich CTCs from the surrounding normal hematopoietic cells. These enrichment methods rely on different properties of CTCs: (a) biological properties (e.g., surface protein expression) and (b) physical properties (e.g., size, density, electric charges, and deformability).

Biological properties are mainly used in immunological procedures with antibodies against either tumor-associated antigens (positive selection) or common leukocytes antigen CD45 (negative selection). Positive enrichment typically attains high cell purity, which depends on antibody specificity. Among the current positive systems, most of the technologies targeted the epithelial cell adhesion molecule (EpCAM) antigen, as the FDA-cleared CELLSEARCH® system which is frequently compared for all new CTC detection methods as the gold standard. However,

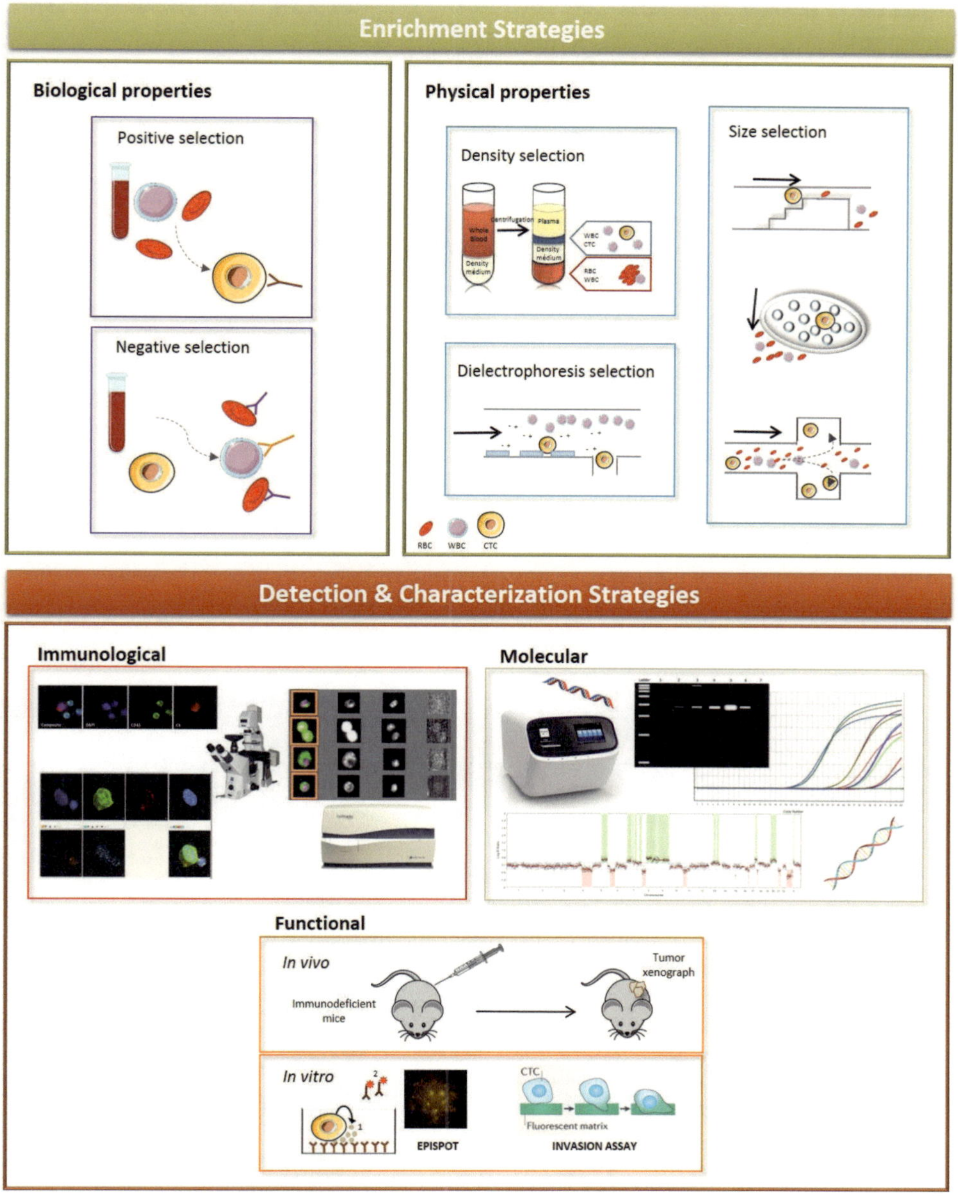

Figure 2.
Strategies for enrichment, detection, and characterization of CTCs.

capturing CTCs lacking EpCAM expression has involved the use of cocktails of antibodies against various other epithelial cell surface antigens (e.g., EGFR, MUC1) or against tissue-specific antigen (e.g., PSA, HER2) and against mesenchymal or stem-cell antigens (e.g., Snail, ALDH1) [7]. Positive selection of CTCs requires an assumption about the unknown nature of CTCs in an individual blood sample. This bias is avoided by negative selection in which the blood sample is depleted of unwanted cells. Indeed, negative enrichment targets and removes background cells, such as leukocytes, using antibodies against CD45 (which is not expressed on carcinomas or other solid tumors) and other leukocyte antigens, to achieve a CTC-enriched sample. Moreover, negative enrichment technologies evade some of the pitfalls of positive enrichment; for example, CTCs are not tagged with a difficult-to-remove antibody, they are not activated or modified via an antibody-protein

SUBCATEGORIES	CRITERIA	TECHNOLOGIES	ADVANTAGES	DISADVANTAGES	COMMERCIAL STATUS
Enrichment methods					
Biological properties					
Positive selection	Immunomagnetic EpCAM	CellSearch®	Robust - Reproducible - Automated	Miss CTCs lacking or with very low expression of EpCAM	Silicon BioSystems-Menarini FDA approved - IVD
		MACS	No centrifugation step		Miltenyi RUO
		MagSwepper			Not commercialized
	Immunomagnetic Antibodies cocktail	AdnaTest	Based on a panel of proteins expressed on CTCs		AdnaGen RUO
	Microfluidic EpCAM	Ephesia Chip		Miss CTCs lacking or with very low expression of EpCAM	Not commercialized
		IsoFlux™	Automated		Fluxion RUO
		CTC-iChip	Combination of biological and physical properties		Not commercialized
	Microfluidic Antibodies cocktail	OncoCEE	Based on cocktail of Abs to capture CTCs		Biocept RUO
	Surface capture	CytoTrapNano™			CytoLumina RUO
	In vivo	CellCollector®	Screen a large volume of blood	Cells can not be removed from the device for the an easy CTC detection	Gilupi CE approved
Negative sélection	Immuno complexes	RosetteSep™ CTC Enrichment	No bias based on protein marker expression nor size of CTCs	Leucocyte contamination	STEMCELL Technologies RUO
	Immunomagnetic	EasySepTM			STEMCELL Technologies RUO
		MACS			Miltenyi RUO
	Microfluidic	CTC-iChip			Not commercialized
Physical properties					
Density selection		Ficoll-Paque®	Fast and cheap No specific equipment needed	Heterogeneity of the density of CTCs	GE Healthcare
		OncoQuick®			Greiner Bio-One
Size based selection	Microfiltration	ISET®	Automated	Heterogeneity of the size and deformability of CTCs	Rarecells RUO
		ScreenCell®			ScreenCell RUO
		CellSieve™			Creatv MicroTech RUO
		Parsortix™	Automated		Angle RUO
	Inertial focusing	VTX-1	Automated		Vortex Biosciences RUO
		ClearCell® FX		Need of RBC lysis before enrichment	Clearbridge BioMedics RUO
Dielectrophoretic charge		ApoStream®		Need a pre enrichment step with ficoll gradient centrifugation	APOCELL RUO
Cluster selection		Cluster-Chip			Not commercialized
Detection / Characterization method					
Immunological					
	Imaging	CellSearch®	Standardized - Robust - Reproducible - Automated	Limited for the protein characterisation - only one channel free	Silicon BioSystems FDA approved - IVD
		IsoFlux Cytation Imager			Fluxion RUO
		CytoTrapNano™	Only one equiment for enrichment and detection step		CytoLumina RUO
		Ikoniscope®	Hight resolution imaging Fish tests associated	Standard microscopy	Ikonisys
		Ariol system	Hight resolution imaging Software for clinical analysis		Leica Biosystems
	FAST technology	FASTCell™	No need of enrichment method before		SRI International
		CytoTrack CT11™	No need of enrichment method before Pick up of cells		CytoTrack
		EPIC	Only lysis of RBC as enrichment step	No equipment avaliable Only the service is commercialized	EPIC Science
	Flow Cytometry	Different kind of cytometer	Sorting capacity for some cytometer	Less sensitive	
	Dielectrophoresis microscopy	DEPArray	Sorting capacity at single cell level		Silicon BioSystems RUO
Nucleic acid based					
	mRNA	screenCell®	Specific panels for each cancer types	No CTC observation	AdnaGen RUO
		Genesis System	High throughput technology Microscopy avaliable for CTC observation		Celsee RUO
	RNA sequencing DNA mutation DNA methylation DNA sequencing	Specific to each lab and research studies			Not commercialized
Functional					
	CTC culture	EPISPOT	Detection of viable CTCs	No possibility to recover CTCs for downstream analysis	Not commercialized
		EPIDROP	Detection of the total number of CTCs plus viables CTCs at the single cell level		Not commercialized
		Vita-Assay™	Detection of viable CTCs		Vitatex RUO
	In vivo	PDX/CDX models	Specific to each lab and research studies		Not commercialized

Table 1.
Advantages, disadvantages, and commercial status of technologies for enrichment, detection, and characterization of CTCs.

interaction, and antibody selection does not bias the subpopulation of CTCs captured. However, these advantages come at the cost of purity, as negative enrichment strategies typically have a much lower purity than positive enrichment [8–10] and require a suitable CTC detection step.

These last years, numerous marker-independent techniques have been developed for CTC isolation and detection. Label-free enrichment process based on physical properties, such as density, size, deformability, and electric charge, have come to avoid molecular bias induced by variability of cell biomarker expression associated with tumor heterogeneity. Mostly used, size and density technologies like microfiltration technologies, based on the precedent that CTCs generally exhibit a larger morphology than leukocytes, or microfluidic devices using inertial focusing to separate CTCs from blood are developed by several companies such as ScreenCell® [11], ISET® [12], CellSieve™ [13, 14], Parsortix™ [15], or Vortex [16]. Such technologies or approaches have the advantages of being less complicated, sometimes rapid, and require minimal equipment. However, some of these approaches may be prone to clogging, and the release of the CTCs into suspension for further analysis is challenging.

2.2 Strategies for CTC detection

After enrichment, the CTC fraction still contains a substantial number of leukocytes, and CTCs need to be specifically identified at the single-cell level by a robust and reproducible method that can distinguish them from normal blood cells.

Immunological technologies are the most frequent methods used for CTC detection using a combination of membrane and/or intracytoplasmic anti-epithelial, anti-mesenchymal, and anti-tissue-specific marker or antitumor-associated antibodies [7]. However, many CTC assays use the same identification step as the CELLSEARCH® system: cells are fluorescently stained for cytokeratins (CK), the common leukocyte antigen CD45, and a nuclear dye (DAPI).

Nucleic acid-based CTC detection methods are the most widely used alternatives to immunological assays to identify CTCs. These techniques identify specific tumor DNA or mRNA to confirm the presence of CTCs indirectly [17]. Detection involves designing specific primers supposedly associated with CTC-specific genes. These genes either code for tissue-, organ-, or tumor-specific proteins or, more specifically, contain known mutations, translocations, or methylation patterns found in cancer cells [18]. These methods have the highest sensitivity but lack specificity, owing to the potential of captured noncancerous cells to generate false-positive signals, thus decreasing the overall accuracy. Considering the genetic heterogeneity of CTCs, multiplex PCR, such as the AdnaTest kit (AdnaGen AG), could overcome this limitation [19, 20].

Furthermore, functional assays that exploit aspects of live cellular activity for CTC detection have the particularity to focus on the discovery of the "metastasis-competent cells." The functional epithelial immunospot (EPISPOT) assay was introduced for in vitro CTC detection and focuses only in viable CTCs [21]. This technology assesses the presence of CTCs based on secretion, shedding, or release of specific proteins during 24–48 h of short-term culture [22]. More recently, Tang et al. described a high-throughput metabolic-based assay for rapid detection of rare metabolically active tumor cells in pleural effusion and peripheral blood of lung cancer patients [23]. In vivo, important information can be obtained by transplantation of patient-derived CTCs into immunodeficient mice: tumors that could grow after xenotransplantation of enriched CTCs have the characteristics of metastasis-initiator cells [8].

2.3 Strategies for CTC characterization

CTCs hold the key to understand the biology of metastasis and provide a biomarker to noninvasively measure the evolution of tumor subclone during treatment and disease progression. Improvements in technologies to yield purer CTC populations make better cellular and molecular investigation. Characterization of CTCs allows better insight into tumor heterogeneity, within most assays, including immunofluorescence, array CGH, next-generation sequencing (NGS) of both DNA and RNA, and fluorescence in situ hybridization.

Protein analyses on single CTCs are currently performed by immunostaining with antibodies directed against protein of interest. Multiple labeling is possible but usually restricted to a few proteins of interest for tumor cell biology and cancer therapy. This may help to identify signaling pathways relevant to metastasis development and treatment responses. In breast cancer patient, the HER2 status of CTCs could be assessed and shows discrepancies with primary tumor status [24, 25]. More recently, immune checkpoint regulators such as programmed death-ligand 1 (PD-L1) have become exciting new therapeutic targets and could be used for *liquid biopsy* in future clinical trials on patients undergoing immune checkpoint blockage [26, 27].

Immunological detection and characterization offer the advantage of allowing isolation of stained CTCs for subsequent molecular characterization. While manual isolation by micromanipulation of CTCs is possible [28], it is rather arduous and time-consuming. An alternative automated single-cell selection device has been therefore developed. The DEPArray™ technology based on a dielectrophoresis strategy by trapping single cells in DEP cages [29] is designed for single-cell recovery of CTCs. Multiple clinical studies have used DEPArray™ to detect and recover single CTCs for subsequent genetic analyses [30–32].

Among single-cell sequencing to identify genomic and transcriptomic characteristics of CTCs, most studies have focused on genomic analyses and carried out whole genome amplifications (WGAs) to increase the amount of DNA, which is subsequently subjected to the analyses of specific mutations and copies number variations using conventional and next-generation sequencing technologies [28, 33, 34]. As an example, CTCs with mutated KRAS genes will escape anti-EGFR therapy, and their early detection might help to guide therapy in individual patients. Besides isolation of single CTCs, a 3–4 log units enrichment step are enough to detect CTCs based on recently developed highly sensitive technologies (e.g., droplet digital PCR) [35].

Another approach is fluorescence in situ hybridization (FISH) analysis of single CTCs identified by immunocytochemistry [36, 37]. Such an immuno-FISH approach can be combined with automated detection of CTCs and might be easier to implement in future clinical diagnostics. Recently, padlock probe technology, which enables in situ analysis of AR-V7 in CTCs, showed that 71% (22 of 31) of CRPC patients had detectable AR-V7 expression ranging from low to high expression [38]. Patients with AR-V7-positive circulating tumor cells (CTCs) have greater benefit of taxane-based chemotherapy than novel hormonal therapies, indicating a treatment-selection biomarker [39, 40].

Finally, these last years, many teams tried to obtain CTC lines by culturing CTCs *ex vivo*. The establishment of in vitro cultures and permanent lines from CTCs has become a challenging task. Indeed, CTC lines could be used to identify proteins and pathways involved in cancer cell stemness and dissemination and also to test new drugs to inhibit metastasis-competent CTCs. *Ex vivo* CTC cultures have been established for breast [41, 42], prostate [43], lung [44], colon [22], and head and neck cancer [45]. To our knowledge, permanent CTC lines have been described only from circulating colon cancer cells: one before (CTC-MCC-41) [22, 46] and eight after the initiation of the anticancer treatment [47].

3. Biology of CTCs

3.1 Epithelial to mesenchymal plasticity

Epithelial to mesenchymal transition (EMT), which is characterized by the downregulation of epithelial proteins and upregulation of mesenchymal proteins, is a complex process that supports the migratory capacity of epithelial tumor cells and is thought to play a crucial role in promoting cancer metastasis. EMT led to increased motility via rearrangements of cellular contact junctions and loss of cell adhesion (i.e., E-cadherin, N-cadherin, claudins), plus epithelial cell morphology through cytoskeleton modification (i.e., cytokeratin, vimentin, fibronectin, etc.) [48]. This invasive phenotype enables cancer cells to pass through the basal membrane and endothelial barriers of blood vessels to reach bloodstream. However, it is still unclear what degree of EMT is needed in tumor cells to attain the circulation.

Despite the wealth of experimental data, the exact role of EMT in cancer patients remains more controversial. Over the past 10 years, the development of sensitive technologies that allow the detection and molecular characterization of CTCs helped to shed new light into the importance of EMT for human tumor cell dissemination [7, 49]. All these data lead now to a new trend, focused on plasticity of tumor cell: epithelial to mesenchymal plasticity (EMP) associated with stemness. This process is today considered as a central actor of the metastatic cascade, providing tumor cells the ability to adapt to the different microenvironments encountered during metastatic spread to colonized organs (i.e., adjacent stroma, blood, newly colonized organs).

CTCs with mesenchymal and stemness features can be attributed in some clinical studies to higher disease stages and metastasis [50–52] and even to therapy response and worse outcome [53–55]. However, the published studies addressing the impact of mesenchymal-like CTCs show heterogeneity with regard to assay specificity, size of cancer and control groups, and endpoint parameters.

To conclude, evaluation of the EMT and stem-cell markers in CTCs may provide information of clinical interest, and using these markers to classify CTCs can elucidate CTC heterogeneity. Nevertheless, studies still suffer from lack of standardized procedures and small sample sizes. Therefore, larger well-designed clinical trials are needed to further illuminate the potential values of EMT markers in CTCs.

3.2 Anoikis resistance

In normal tissue, adhesion to appropriate extracellular matrix proteins is essential for survival. Loss of this adhesion induces cell death which has been termed "anoikis." Anoikis is a physiologically relevant process for tissue homeostasis and development because it prevents detached epithelial cells from colonizing elsewhere, thereby inhibiting dysplastic cell growth or attachment to an inappropriate matrix [56]. Dysregulation of anoikis, such as anoikis resistance, is a critical mechanism in tumor metastasis. If cells acquire oncogenic signals that are able to overcome this machinery, they gain the ability to survive outside their normal environment in the absence of adhesion to the extracellular matrix. The tumor cells that acquire anoikis resistance can survive detachment from their primary site, traveling through the circulatory and lymphatic systems to disseminate to ectopic locations [57]. Different studies have shown that the death receptor pathway of caspase activation mediates anoikis; thus, defects in this pathway such as overexpression of the caspase-8 inhibitor FLIP can turn cell resistant to anoikis. Similarly, resistance to anoikis can be conferred by roadblocks in the mitochondrial pathway, such as overexpression of the Bcl-2 family of anti-apoptotic proteins [57].

The investigation of molecular mechanisms involved in cancer cell survival while they are leaving the adherent microenvironment of the tumor to the circulatory system is important to understand the process by which cancer can spread to distant organs, as well as to design new therapeutics to inhibit the spread of the disease.

3.3 Escape to the immune system

Once in the bloodstream, CTCs face several natural obstacles that hinder the metastatic process. One of the main obstacles that CTCs face in the blood is the attack of the immune system. Lots of work was done to understand mechanisms involve in the battle between the immune system's capabilities to fight cancer and the immune-suppressive processes that promote tumor growth. Several biomarkers showed up from this work; for example, in colorectal cancer, immune escape was observed by the upregulation of CD47, a "don't eat me signal" that prevents CTCs from macrophage and dendritic cell attack [58]. The most clinically advanced biomarkers are the programmed death-1 (PD-1) and its ligand (PD-L1). PD-L1 expressed in tumors has been highlighted to function as a key component of the cancer-immunity cycle by preventing the immune system from destroying cancer cells. PD-1 receptor is a surface protein expressed on activated T-cells, and its ligand PD-L1 is expressed on the surface of antigen-presenting cells. The formation of the PD-1/PD-L1 complex induces a strong inhibitory signal in the T-cell, which leads to a reduction of cytokine production and a suppression of T-cell proliferation [59]: the immune system is misled by the cancer cells expressing PD-L1 and does not destroy them. That understanding led to the development of immune checkpoint inhibitor therapies, antibodies against both PD-1 and PD-L1, and remarkable clinical responses which have been seen in several different malignancies including, but not limited to, melanoma, lung, kidney, and bladder cancers [59].

However, CTCs can use several mechanisms to survive in the circulatory system. For example, these cells can couple to reactive platelets. Several hypotheses propose that the surface coating of platelets may serve as a shield against immune assault or that platelets may load the major histocompatibility complex to CTCs to imitate host cells and therefore avoid immune surveillance [60]. The aggregation of CTCs with platelets, stromal fibroblasts, and leukocytes leads to the formation of floating complexes and increases the survival of CTCs in the bloodstream by avoiding anoikis and killing by immune cells [61]. In addition, the vascular endothelial growth factor (VEGF), secreted by platelets, is able to affect the maturation of dendritic cells that play a key role in antigen presentation [62].

3.4 CTC microemboli

An alternative mechanism for metastasis has emerged from recent studies, the collective migration of tumor cells by clusters of CTCs. CTC clusters are defined as groups of tumor cells (more than two or three cells, varied among studies) that travel together in the bloodstream. Thus, in the blood circulation, CTCs can be found both as single tumor cells and clusters of tumor cells in patients with an advanced stage of the cancer. Study using mouse models with tagged mammary tumors demonstrates that these clusters arise from oligoclonal tumor cell groupings and not from intravascular aggregation events [63]. Moreover, CTC clusters have 23- to 50-fold increased metastatic potential. Even fewer in number, clusters of CTCs possess much higher metastatic potential than individual CTCs.

Patients with CTC microemboli or clusters in their bloodstream have significantly worse overall and progression-free survival than those with only individually migrating single CTCs [63]. The prognostic value of CTC clusters can be estimated by clinical observations.

Current studies have partially elucidated the reasons for CTC clusters to have higher potential of metastasis. First, tumor cells within CTC clusters showed prolonged survival and decreased apoptosis [64]. Second, the physical specialty of CTC clusters allows for a greater likelihood of it residing in distant organs. Microvasculature of viscera can retain large CTCs; thus, it can retain CTC clusters more easily [65].

4. Clinical relevance of CTCs

Despite many clinical validation studies, CTCs have not been included yet into the clinical guidelines (e.g., ASCO guidelines at http://www.asco.org/practice-guidelines/quality-guidelines/guidelines). Although CTC enumeration can improve current tumor staging and contribute to the early assessment of therapy effects, the clinical utility of CTCs remains to be addressed in interventional studies (i.e., its capacity to decide adopting or to rejecting a therapeutic action).

In this chapter, we highlighted the clinical relevance of CTCs in breast, prostate, colon, and lung cancer. **Figure 3** illustrates how CTCs as *liquid biopsy* can guide clinicians to personalized medicine.

4.1 Breast

More advanced studies, regarding clinical utility of CTCs, are related to metastatic breast cancer (MBC). Sequential CTC enumeration has been shown in a large multicenter prognostic study to be superior to conventional serum protein markers (CA-15-3, CEA) for early detection of therapy failure in MBC [5]. However, in the

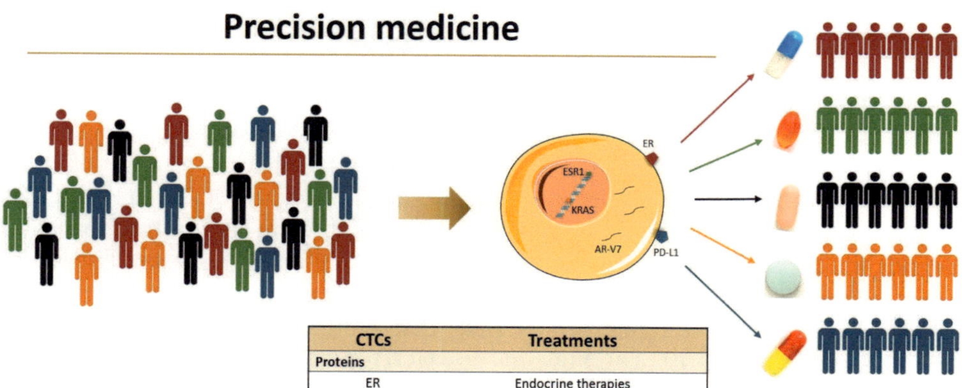

CTCs	Treatments
Proteins	
ER	Endocrine therapies
Her2-neu	Her2 targeted therapies
PD-L1	Immunotherapies - PD-L1 inhibitors
ARNm	
AR-V7	Resistance to endocrine therapies
DNA mutation	
KRAS	EGFR targeted therapies
BRAF	BRAF targeted therapies
PI3K	PI3K pathway inhibitors Resistance to EGFR therapies
ESR1	Resistance to endocrine therapies
T790M	Third generation of Tyrosine Kinase Inhibitors

Figure 3.
CTC as liquid biopsy for precision medicine.

interventional trial SWOG 0500 (NCT00382018), although the prognostic significance of CTCs was confirmed, the CTC-driven switch to an alternate cytotoxic therapy was not effective in prolonging overall survival for MBC patients with persistently increased CTCs after 21 days of therapy [66]. The inconvenient of these kinds of interventional biomarker-driven studies is the fact that the result is dependent of the therapy efficacy. This strategy can only work if there is an efficient therapy for the cohort identified by the test.

Another promising approach is the stratification of patients to chemotherapy or hormonal therapy based on CTC enumeration like in the interventional STIC CTC METABREAST clinical trial (NCT0 1,710,605) for MBC patients [67]. Besides CTC enumeration, stratification based on CTC phenotype might become also an important strategy. Stratification of MBC patients based on HER2 status of CTCs is currently tested in the DETECT III trial [67].

Other possible uses for CTC detection include prognostication in early stage patients, identifying patients requiring adjuvant therapy. The SUCCESS study provides strong evidence of the prognostic relevance of CTCs in early breast cancer before and after adjuvant chemotherapy in a large patient cohort [68]. This study outlines the potential of the CTC analysis at primary diagnostic to evaluate individual risk and points that they may use for treatment management in early stage of cancer. These data have been confirmed by Bidard et al. who conducted a meta-analysis in nonmetastatic breast cancer patients treated by neoadjuvant chemotherapy (NCT) to assess the clinical validity of CTC detection as a prognostic marker [69]. They showed that CTC count is an independent and quantitative prognostic factor in early breast cancer patients treated by NCT. *Liquid biopsy* complements current prognostic models based on tumor characteristics and response to therapy. Moreover, Trapp et al. demonstrated recently that the presence of CTCs 2 years after chemotherapy was associated with decreased OS and DFS. Based on these results, active individualized surveillance strategies for breast cancer survivors based on biomarkers should be reconsidered [70].

4.2 Prostate

For men with metastatic castration-resistant prostate cancer (mCRPC), the CELLSEARCH® system method for CTCs enumeration is the only FDA-cleared CTC test available clinically. The CTC count has been shown to provide prognostic value and was associated with treatment response in mCRPC patients, in several recent studies [4, 71–73], indicating a clear value as a patient-level indicator of survival. However, despite increasing evidence that CTCs could be used to monitor disease progression in mCRPC [18], CTC use is still limited to clinical trials in academic centers. Clinical utility of CTCs, reflecting the ability of this test to favorably change outcomes, is still an unmet clinical need in prostate cancer [74]. The first interventional clinical trial in prostate cancer that will show the clinical utility of CTCs should start in 2019 (TACTIK project—NCT03101046).

Moreover new data suggest that CTCs may harbor genetic information (such as the androgen receptor splice variant 7, AR-V7) relevant to changing clinical management and predicting treatment sensitivity or resistance to cancer therapies such as enzalutamide, abiraterone, and taxane-based chemotherapies [39].

Regarding nonmetastatic cancer patient, a recent European TRANSCAN study CTC-SCAN investigated the feasibility of detecting CTCs in nonmetastatic high-risk prostate cancer (PCa) patients by combining the CELLSEARCH® platform, the in vivo CellCollector® capture system, and the EPISPOT assay. The observed correlation, with established risk factors and the persistence of CTCs 3 months after surgery, suggested a potential clinical relevance of CTCs as markers of minimal residual disease (MRD) in PCa [75]. CTC-based liquid biopsies have the potential to monitor MRD in patients with nonmetastatic prostate cancer although follow-up evaluations are now required to assess how to provide independent prognostic information. A new European project (Transcan—PROLIPSY) will assess whether CTCs in combination with exosomes and ctDNA as noninvasive *liquid biopsy* allow the diagnosis of prostate cancer and the evaluation of its aggressiveness.

4.3 Colon

In 2008, Cohen et al. demonstrated the independent prognostic and predictive value of CTCs for patients initiating chemotherapy for metastatic colorectal cancer (mCRC) [2]. Since this first publication defining a cutoff of three CTCs, different meta-analyses have confirmed that baseline levels of CTC count is an important prognostic factor for PFS and OS in patients with mCRC [76–78].

Despite the strong evidence of a prognostic significance of CTC count, there is no solid evidence demonstrating the interest of CTC count for therapeutic strategy, and this biomarker is rarely used in the management of patients with mCRC. However, patients with high CTC counts recruited in a phase II study could benefit from a more intense chemotherapeutic regimen [79]. These preliminary data require validation in randomized trials. Moreover, Lalmahomed et al. failed to show a prognostic effect of CTCs for early relapse after the resection of colorectal liver metastases [80].

4.4 Lung

The role of CTCs in non-small cell lung cancer (NSCLC) has been addressed in several clinical trials. More specifically, the prediction of the outcome of patients with early and advanced NSCLC based on the CTC enumeration has been explored. The CTC count with the CELLSEARCH® system in advanced NSCLC patients who received standard chemotherapy was associated with a shorter PFS and OS, but standardize cutoff could not be observed [81–84]. Furthermore, analysis of CTCs from patients with metastatic NSCLC identified the expected EGFR-activating mutation in CTCs from 11 of 12 patients (92%) and in matched free plasma DNA from 4 of 12 patients (33%) [85]. The T790 M mutation, which confers drug resistance, was revealed in CTCs from patients who had received tyrosine kinase inhibitors, suggesting the strong potential gain of noninvasive liquid biopsy. Moreover, serial increases in CTC counts were associated with tumor progression, with the emergence of additional EGFR mutations in some cases. Recently, KRAS and EGFR mutations, relevant for treatment decisions, could be detected in CTCs and in the corresponding primary tumors of the same patients [86].

5. Other circulating biomarkers as *liquid biopsy*

Even if the term "*liquid biopsy*" was originally used for CTC analysis, currently, it includes all different circulating biomarkers like circulating cell-free DNA (cfDNA), microRNA (miRNA), and exosomes that are shed into the bloodstream by tumors and/or metastatic deposits, as well as tumor-educated platelets which are described to have a role in tumor metastasis. Like CTCs, all these other circulating biomarkers need to be validated in clinical trials. **Table 2** summarizes observational and interventional clinical trials on breast, lung, prostate, and colorectal cancer registered in *clinical.gov* (A) and the applications (B).

A.

Biomarkers / Cancer types	CTCs		Circulating DNA		Exosomes		microRNA		TEPs
Breast	**816**		**43**		**5**		**6**		**0**
Interventional		755		24		4		3	
Observational		61		19		1		3	
Lung	**792**		**77**		**9**		**0**		**0**
Interventional		733		25		5		-	
Observational		59		52		4		-	
Prostate	**515**		**25**		**8**		**2**		**0**
Interventional		475		10		2		0	
Observational		40		15		6		2	
Colorectal	**528**		**55**		**4**		**2**		**0**
Interventional		474		18		3		1	
Observational		54		37		1		1	

B.

CLINICAL TRIALS	CTCs	Circulating DNA	Exosomes	microRNA	TEPs
Cancer screening	X	X			
Pronostic value	X	X	X	X	
Treatment response	X	X	X	X	
Protein marker characterization	X				
Mutation analysis	X	X			
Biologial process understanding	X	X	X	X	

Table 2.
(A) Number of observational and interventional clinical trials (clinical.gov) involving ***liquid biopsy*** *in the main cancer types and (B) the applications of each circulating biomarkers (CTCs, circulating DNA, exosomes, microRNA, and TEPs).*

5.1 Circulating tumor DNA

Apoptotic and necrotic tumor cells are known to discharge cell-free nucleic acid fragments into the bloodstream of cancer patients. Although most circulating DNA is believed to originate from nonmalignant cells, an increased level of cfDNA was observed in blood of patients with late stage cancer [87]. Among the pool of total cfDNA, there is circulating tumor DNA (ctDNA) which cannot be specifically isolated from the total pool but can be detected by tumor-specific mutations [88]. In general, cfDNA can be analyzed from plasma by targeted or untargeted approaches. The targeted approaches involve the detection of known genetic changes, e.g., "druggable" mutations, with impact on therapy decisions [89]. The interest of cfDNA was demonstrated by Douillard et al. [90] by determining the EGFR mutational status in NSCLC and can represent a substitute for tissue biopsies when these are not available. Moreover, in 2016, the detection of EGFR gene mutations in cfDNA using the cobas EGFR Mutation Test v2 achieved FDA approval as a companion diagnostic for erlotinib, becoming the first blood-based biopsy test approved for implementation in clinical decisions [91].

However, despite the evidence of potential clinical utility and even if it has been recommended (e.g., by the FDA) that the blood could be analyzed first to reduce the number of invasive biopsies in cancer patients, the lower sensitivity of ctDNA analyses prevents its use in clinical management for the moment, and the primary tumor analysis still remains the gold standard in NSCLC diagnostics of EGFR mutations.

5.2 MicroRNAs

MicroRNAs (miRNAs, miR-x), consisting in approximately 22 nucleotides, represent another potential blood biomarker in oncology. These noncoding small RNAs are master regulators of genic expression and consequently of many cellular processes. Alterations in the expression of microRNA genes have been shown to play in important role in human malignancies. These alterations can be caused by a variety of mechanisms, including deletions, amplifications, or mutations involving microRNA loci, by epigenetic silencing or by dysregulation of transcription factors targeting specific microRNAs [92]. The three major detection techniques for circulating cell-free miRNA (cfmiRNA) analysis, following RNA extraction, comprise quantitative RT-PCR, microarray analyses, and deep sequencing. The assessment of cfmiRNA has been suggested for early diagnosis, prognosis, therapy monitoring, and therapeutic response prediction in different cancer types (e.g., lung, breast, colon, prostate, and ovary cancers and melanoma), as reviewed by Armand-Labit and Pradines [93].

5.3 Exosomes

Tumor and normal cells are known to release microvesicles such as exosomes (40–150 nm) into the circulation, discharging cellular content. Currently, one challenge for the analyses of circulating cell-free nucleic acids in blood is their instability. Thus, due to their protective environment, the exosomes represent a valuable source for analysis of proteins, DNA, RNA, miRNA, lipids, and metabolites [94]. Ultracentrifugation, density-based separation, or immune-affinity capture using magnetic beads coated with anti-EpCAM antibodies can be used to isolate exosomes [95]. They are important regulators of the cellular niche, and their altered characteristics in many diseases, such as cancer, suggest their importance for diagnostic and therapeutic applications and as drug delivery vehicles. Hoshino et al. demonstrated that the composition of exosomal integrins could predict organ-specific metastasis and that tumor-derived exosomes participate in preparing the pre-metastatic niche [96]. Correspondingly, the same group shows that a pro-metastatic phenotype of bone marrow progenitor cells is promoted by education through melanoma exosomes [97].

5.4 Tumor-educated platelets

A new emerging class of components for *liquid biopsy* is tumor-educated platelets (TEPs). These anucleated blood cells (second most abundant cell type in circulation) could be educated by tumor cells by the transfer of tumor-associated biomolecules, mostly RNA. Platelets are isolated by centrifugation and RNA can be subjected to RT-PCR [45]. Performing mRNA sequencing on TEPs, Best and his colleagues showed that cancer patients with different tumor types could be discriminated from healthy individuals with 96% accuracy and that the primary tumor was correctly located with a precision of 71% [98]. Studies have shown that platelet count and platelet size can already provide clinically relevant information about the presence of cancer [99]. High platelet count is associated with increased mortality in a variety of cancers.

Furthermore, biomarkers (MET or HER2 expression/KRAS, EGFR, and PIK3CA mutations) were identified in surrogate TEP mRNA profiles, which might be tested in future studies as potential predictors for targeted therapies. Recently, Diem et al. showed that elevated pretreatment platelet-to-lymphocyte ratios correlate with a reduced response rate to nivolumab anti-PD-L1 immunotherapy in NSCLC [100], indicating that circulating platelets may enhance a pro-tumorigenic effect in the presence of an antitumor immune response.

6. Conclusion

CTC as *liquid biopsy* represents a promising approach for personalized treatment in oncology. Lots of efforts have been made to overcome technical challenges for enrichment, detection, and characterization of these tumor cells. Nevertheless, low number (or even absence) of CTCs can weaken the reliability of CTC-based assays in some patients with current detection techniques. This points the need for further technological advances and procedure standardization. To introduce CTC tests into clinical trials, an intense validation of the technical aspects of the applied assays is currently executed in Europe by the EU-funded CANCER-ID network (www.cancer-id.eu) that will be continued by the European *Liquid Biopsy* Society (ELBS).

Additionally, an extensive work has been made to understand biological processes of cancer dissemination and metastasis, underlying different aspects for CTCs survival in bloodstream. This knowledge could improve pharmaceutical drug researches and therapeutic strategies for better clinical management of cancer patients.

Beside CTC analysis several other circulating biomarkers are under investigations and demonstrate real valuable data. It is now well accepted that there is not a perfect unique biomarker and that combining different circulating biomarkers can bring a huge benefit for precision medicine for cancer patients.

In conclusion, *liquid biopsy* diagnostics might help to focus the current cancer screening modalities, which would reduce side and healthcare costs. However, despite promising first results and the enormous interest by diagnostic companies and the public press, disease monitoring and early detection of cancer face serious challenges of both sensitivity and specificity.

Acknowledgements

The authors received support from (1) the National Institute of Cancer (INCa, http://www.e-cancer.fr), (2) CANCER-ID, an Innovative Medicines Initiative Joint Undertaking under grant agreement no. 115749, resources of which are composed of financial contribution from the European Union's Seventh Framework Program (FP7/2007-2013) (www.cancer-id.eu) and EFPIA companies' in-kind contribution, (3) ARC Foundation, (4) Ligue contre le cancer, and (5) the ELBA—Innovative Training Networks (ITN) H2020—European Liquid Biopsies Academy project—Toward widespread clinical application of blood-based diagnostic tools. H2020-MSCA-ITN-2017 (http://elba.uni-plovdiv.bg).

Author details

Laure Cayrefourcq* and Catherine Alix-Panabières
Laboratory of Rare Human Circulating Cells (LCCRH), University Medical Centre of Montpellier, Montpellier, France

*Address all correspondence to: l-cayrefourcq@chu-montpellier.fr

References

[1] Pantel K, Alix-Panabieres C. Circulating tumour cells in cancer patients: Challenges and perspectives. Trends in Molecular Medicine. 2010;**16**(9):398-406

[2] Cohen SJ et al. Relationship of circulating tumor cells to tumor response, progression-free survival, and overall survival in patients with metastatic colorectal cancer. Journal of Clinical Oncology. 2008;**26**(19):3213-3221

[3] Cristofanilli M et al. Circulating tumor cells, disease progression, and survival in metastatic breast cancer. New England Journal of Medicine. 2004;**351**(8):781-791

[4] de Bono JS et al. Circulating tumor cells predict survival benefit from treatment in metastatic castration-resistant prostate cancer. Clinical Cancer Research. 2008;**14**(19):6302-6309

[5] Bidard FC et al. Clinical validity of circulating tumour cells in patients with metastatic breast cancer: A pooled analysis of individual patient data. The Lancet Oncology. 2014;**15**(4):406-414

[6] Zhang L et al. Meta-analysis of the prognostic value of circulating tumor cells in breast cancer. Clinical Cancer Research. 2012;**18**(20):5701-5710

[7] Alix-Panabieres C, Mader S, Pantel K. Epithelial-mesenchymal plasticity in circulating tumor cells. Journal of Molecular Medicine. 2017;**95**(2):133-142

[8] Baccelli I et al. Identification of a population of blood circulating tumor cells from breast cancer patients that initiates metastasis in a xenograft assay. Nature Biotechnology. 2013;**31**(6):539-544

[9] Yang L et al. Optimization of an enrichment process for circulating tumor cells from the blood of head and neck cancer patients through depletion of normal cells. Biotechnology and Bioengineering. 2009;**102**(2):521-534

[10] Lara O et al. Enrichment of rare cancer cells through depletion of normal cells using density and flow-through, immunomagnetic cell separation. Experimental Hematology. 2004;**32**(10):891-904

[11] Desitter I et al. A new device for rapid isolation by size and characterization of rare circulating tumor cells. Anticancer Research. 2011;**31**(2):427-441

[12] Farace F et al. A direct comparison of CellSearch and ISET for circulating tumour-cell detection in patients with metastatic carcinomas. British Journal of Cancer. 2011;**105**(6):847-853

[13] Adams DL et al. The systematic study of circulating tumor cell isolation using lithographic microfilters. RSC Advances. 2014;**9**:4334-4342

[14] Adams DL et al. Cytometric characterization of circulating tumor cells captured by microfiltration and their correlation to the CellSearch((R)) CTC test. Cytometry. Part A. 2015;**87**(2):137-144

[15] Xu L et al. Optimization and evaluation of a novel size based circulating tumor cell isolation system. PLoS One. 2015;**10**(9):e0138032

[16] Lemaire CA et al. Fast and label-free isolation of circulating tumor cells from blood: From a research microfluidic platform to an automated fluidic instrument, VTX-1 liquid biopsy system. SLAS Technology. 2018;**23**(1):16-29

[17] Esmaeilsabzali H et al. Detection and isolation of circulating tumor cells:

Principles and methods. Biotechnology Advances. 2013;**31**(7):1063-1084

[18] Riethdorf S, Wikman H, Pantel K. Review: Biological relevance of disseminated tumor cells in cancer patients. International Journal of Cancer. 2008;**123**(9):1991-2006

[19] Chebouti I et al. ERCC1-expressing circulating tumor cells as a potential diagnostic tool for monitoring response to platinum-based chemotherapy and for predicting post-therapeutic outcome of ovarian cancer. Oncotarget. 2017;**8**(15):24303-24313

[20] Fehm T et al. Detection and characterization of circulating tumor cells in blood of primary breast cancer patients by RT-PCR and comparison to status of bone marrow disseminated cells. Breast Cancer Research. 2009;**11**(4):R59

[21] Soler A et al. EpCAM-independent enrichment and detection of viable circulating tumor cells using the EPISPOT assay. Methods in Molecular Biology. 2017;**1634**:263-276

[22] Cayrefourcq L et al. Establishment and characterization of a cell line from human circulating colon cancer cells. Cancer Research. 2015;**75**(5): 892-901

[23] Tang Y et al. High-throughput screening of rare metabolically active tumor cells in pleural effusion and peripheral blood of lung cancer patients. Proceedings of the National Academy of Sciences USA. 4 Apr 2017;**114**(14):E2983. DOI: 10.1073/pnas.1703650114. Epub 2017 Mar 27

[24] Riethdorf S et al. Detection and HER2 expression of circulating tumor cells: Prospective monitoring in breast cancer patients treated in the neoadjuvant GeparQuattro trial. Clinical Cancer Research. 2010;**16**(9): 2634-2645

[25] Jaeger BAS et al. The HER2 phenotype of circulating tumor cells in HER2-positive early breast cancer: A translational research project of a prospective randomized phase III trial. PLoS One. 2017;**12**(6):e0173593

[26] Topalian SL, Drake CG, Pardoll DM. Immune checkpoint blockade: A common denominator approach to cancer therapy. Cancer Cell. 2015;**27**(4):450-461

[27] Mazel M et al. Frequent expression of PD-L1 on circulating breast cancer cells. Molecular Oncology. 2015;**9**(9):1773-1782

[28] Heitzer E et al. Complex tumor genomes inferred from single circulating tumor cells by array-CGH and next-generation sequencing. Cancer Research. 2013;**73**(10): 2965-2975

[29] Abonnenc M et al. Programmable interactions of functionalized single bioparticles in a dielectrophoresis-based microarray chip. Analytical Chemistry. 2013;**85**(17):8219-8224

[30] Fabbri F et al. Detection and recovery of circulating colon cancer cells using a dielectrophoresis-based device: KRAS mutation status in pure CTCs. Cancer Letters. 2013;**335**(1): 225-231

[31] Mesquita B et al. Molecular analysis of single circulating tumour cells following long-term storage of clinical samples. Molecular Oncology. 2017;**11**(12):1687-1697

[32] Paolillo C et al. Detection of activating Estrogen receptor gene (ESR1) mutations in single circulating tumor cells. Clinical Cancer Research. 2017;**23**(20):6086-6093

[33] Paoletti C et al. Comprehensive mutation and copy number profiling in archived circulating breast cancer tumor cells documents heterogeneous

resistance mechanisms. Cancer Research. 2018;**78**(4):1110-1122

[34] Sho S et al. Precision oncology using a limited number of cells: Optimization of whole genome amplification products for sequencing applications. BMC Cancer. 2017;**17**(1):457

[35] Denis JA, Lacorte JM. Detection of RAS mutations in circulating tumor cells: Applications in colorectal cancer and prospects. Annales de Biologie Clinique. 2017;**75**(6):607-618

[36] Obermayr E et al. Circulating tumor cells: Potential markers of minimal residual disease in ovarian cancer? A study of the OVCAD consortium. Oncotarget. 2017;**8**(63):106415-106428

[37] Podolak J et al. Androgen receptor amplification is concordant between circulating tumor cells and biopsies from men undergoing treatment for metastatic castration resistant prostate cancer. Oncotarget. 2017;**8**(42):71447-71455

[38] El-Heliebi A et al. In situ detection and quantification of AR-V7, AR-FL, PSA, and KRAS point mutations in circulating tumor cells. Clinical Chemistry. 2018;**64**(3):536-546

[39] Antonarakis ES et al. Androgen receptor splice variant 7 and efficacy of taxane chemotherapy in patients with metastatic castration-resistant prostate cancer. JAMA Oncology. 2015;**1**(5):582-591

[40] Scher HI et al. Association of ar-v7 on circulating tumor cells as a treatment-specific biomarker with outcomes and survival in castration-resistant prostate cancer. JAMA Oncology. 2016;**2**(11):1441-1449

[41] Zhang L et al. The identification and characterization of breast cancer CTCs competent for brain metastasis. Science Translational Medicine. 2013;**5**(180):180ra48

[42] Yu M et al. Circulating tumor cells: Approaches to isolation and characterization. The Journal of Cell Biology. 2011;**192**(3):373-382

[43] Gao D et al. Organoid cultures derived from patients with advanced prostate cancer. Cell. 2014;**159**(1):176-187

[44] Zhang Z et al. Expansion of CTCs from early stage lung cancer patients using a microfluidic co-culture model. Oncotarget. 2014;**5**(23):12383-12397

[45] Kulasinghe A et al. Short term ex-vivo expansion of circulating head and neck tumour cells. Oncotarget. 2016;**7**(37):60101-60109

[46] Alix-Panabieres C et al. Molecular portrait of metastasis-competent circulating tumor cells in colon cancer reveals the crucial role of genes regulating energy metabolism and DNA repair. Clinical Chemistry. 2017;**63**(3):700-713

[47] Soler A et al. Autologous cell lines from circulating colon cancer cells captured from sequential liquid biopsies as model to study therapy-driven tumor changes. Scientific Reports. 2018;**8**(1):15931

[48] Nieto MA et al. EMT: 2016. Cell. 2016;**166**(1):21-45

[49] Jie XX, Zhang XY, Xu CJ. Epithelial-to-mesenchymal transition, circulating tumor cells and cancer metastasis: Mechanisms and clinical applications. Oncotarget. 2017;**8**(46): 81558-81571

[50] Kallergi G et al. Epithelial to mesenchymal transition markers expressed in circulating tumour cells of early and metastatic breast cancer

patients. Breast Cancer Research. 2011;**13**(3):R59

[51] Papadaki MA et al. Co-expression of putative stemness and epithelial-to-mesenchymal transition markers on single circulating tumour cells from patients with early and metastatic breast cancer. BMC Cancer. 2014;**14**:651

[52] Alonso-Alconada L et al. Molecular profiling of circulating tumor cells links plasticity to the metastatic process in endometrial cancer. Molecular Cancer. 2014;**13**:223

[53] Chang K et al. Combination of circulating tumor cell enumeration and tumor marker detection in predicting prognosis and treatment effect in metastatic castration-resistant prostate cancer. Oncotarget. 2015;**6**(39):41825-41836

[54] Polioudaki H et al. Variable expression levels of keratin and vimentin reveal differential EMT status of circulating tumor cells and correlation with clinical characteristics and outcome of patients with metastatic breast cancer. BMC Cancer. 2015;**15**:399

[55] Zhao R et al. Expression and clinical relevance of epithelial and mesenchymal markers in circulating tumor cells from colorectal cancer. Oncotarget. 2017;**8**(6):9293-9302

[56] Kim YN et al. Anoikis resistance: An essential prerequisite for tumor metastasis. International Journal of Cell Biology. 2012;**2012**:306879

[57] Simpson CD, Anyiwe K, Schimmer AD. Anoikis resistance and tumor metastasis. Cancer Letters. 2008;**272**(2):177-185

[58] Steinert G et al. Immune escape and survival mechanisms in circulating tumor cells of colorectal cancer. Cancer Research. 2014;**74**(6):1694-1704

[59] Ohaegbulam KC et al. Human cancer immunotherapy with antibodies to the PD-1 and PD-L1 pathway. Trends in Molecular Medicine. 2015;**21**(1):24-33

[60] Lou X-L et al. Interaction between circulating cancer cells and platelets: Clinical implication. Chinese Journal of Cancer Research. 2015;**27**(5):450-460

[61] Hong Y, Fang F, Zhang Q. Circulating tumor cell clusters: What we know and what we expect (review). International Journal of Oncology. 2016;**49**(6):2206-2216

[62] Abbott NJ, Rönnbäck L, Hansson E. Astrocyte–endothelial interactions at the blood–brain barrier. Nature Reviews Neuroscience. 2006;**7**:41

[63] Aceto N et al. Circulating tumor cell clusters are oligoclonal precursors of breast cancer metastasis. Cell. 2014;**158**(5):1110-1122

[64] Hou JM et al. Circulating tumor cells as a window on metastasis biology in lung cancer. The American Journal of Pathology. 2011;**178**(3):989-996

[65] Peeters DJ et al. Circulating tumour cells and lung microvascular tumour cell retention in patients with metastatic breast and cervical cancer. Cancer Letters. 2015;**356**(2 Pt B):872-879

[66] Smerage JB et al. Circulating tumor cells and response to chemotherapy in metastatic breast cancer: SWOG S0500. Journal of Clinical Oncology. 2014;**32**(31):3483-3489

[67] Bidard FC et al. Clinical application of circulating tumor cells in breast cancer: Overview of the current interventional trials. Cancer Metastasis Reviews. 2013;**32**(1-2):179-188

[68] Rack B et al. Circulating tumor cells predict survival in early average-to-high risk breast cancer patients. JNCI Journal of the National Cancer Institute. 2014;**106**(5):dju066

[69] Bidard FC et al. Circulating tumor cells in breast cancer patients treated by neoadjuvant chemotherapy: A meta-analysis. Journal of the National Cancer Institute. 2018;**110**(6):560-567

[70] Trapp E et al. Presence of circulating tumor cells in high-risk early breast cancer during follow-up and prognosis. Journal of the National Cancer Institute. 11 Oct 2018. DOI: 10.1093/jnci/djy152. [Epub ahead of print]

[71] Scher HI et al. Circulating tumor cell biomarker panel as an individual-level surrogate for survival in metastatic castration-resistant prostate cancer. Journal of Clinical Oncology. 2015;**33**(12):1348-1355

[72] Scher HI et al. Circulating tumor cell number as a prognostic marker in progressive castration-resistant prostate cancer: Use in clinical practice and clinical trials. The Lancet Oncology. 2009;**10**(3):233-239

[73] Scher HI et al. Circulating tumour cells as prognostic markers in progressive, castration-resistant prostate cancer: A reanalysis of IMMC38 trial data. The Lancet Oncology. 2009;**10**(3):233-239

[74] Lorente D et al. Interrogating metastatic prostate cancer treatment switch decisions: A multi-institutional survey. European Urology Focus. 2018;**4**(2):235-244

[75] Kuske A et al. Improved detection of circulating tumor cells in non-metastatic high-risk prostate cancer patients. Scientific Reports. 2016;**6**:39736

[76] Groot Koerkamp B et al. Circulating tumor cells and prognosis of patients with Resectable colorectal liver metastases or widespread metastatic colorectal cancer: A meta-analysis. Annals of Surgical Oncology. 2013;**20**(7):2156-2165

[77] Huang X et al. Meta-analysis of the prognostic value of circulating tumor cells detected with the CellSearch System in colorectal cancer. BMC Cancer. 2015;**15**:202

[78] Rahbari NN et al. Meta-analysis shows that detection of circulating tumor cells indicates poor prognosis in patients with colorectal cancer. Gastroenterology. 2010;**138**(5):1714-1726

[79] Krebs MG et al. Circulating tumor cell enumeration in a phase II trial of a four-drug regimen in advanced colorectal cancer. Clinical Colorectal Cancer. 2015;**14**(2):115-122.e1-2

[80] Lalmahomed ZS et al. Prognostic value of circulating tumour cells for early recurrence after resection of colorectal liver metastases. British Journal of Cancer. 2015;**112**(3): 556-561

[81] Hirose T et al. Relationship of circulating tumor cells to the effectiveness of cytotoxic chemotherapy in patients with metastatic non-small-cell lung cancer. Oncology Research Featuring Preclinical and Clinical Cancer Therapeutics. 2012;**20**(2-3):131-137

[82] Krebs MG et al. Evaluation and prognostic significance of circulating tumor cells in patients with non-small-cell lung cancer. Journal of Clinical Oncology. 2011;**29**(12): 1556-1563

[83] Muinelo-Romay L et al. Evaluation of circulating tumor cells and related events as prognostic factors and surrogate biomarkers in advanced NSCLC patients receiving first-line systemic treatment. Cancers. 2014;**6**(1):153

[84] Xu YH, Zhou J, Pan XF. Detecting circulating tumor cells in patients with advanced non-small cell lung cancer. Genetics and Molecular Research. 2015;**14**(3):10352-10358

[85] Maheswaran S et al. Detection of mutations in EGFR in circulating lung-cancer cells. The New England Journal of Medicine. 2008;**359**(4):366-377

[86] Gorges TM et al. Enumeration and molecular characterization of tumor cells in lung cancer patients using a novel in vivo device for capturing circulating tumor cells. Clinical Cancer Research. 2016;**22**(9):2197-2206

[87] Leon SA et al. Free DNA in the serum of cancer patients and the effect of therapy. Cancer Research. 1977;**37**(3):646-650

[88] Pantel K, Alix-Panabières C. Real-time liquid biopsy in cancer patients: Fact or fiction? Cancer Research. 2013;**73**(21):6384-6388

[89] Heitzer E, Ulz P, Geigl JB. Circulating tumor DNA as a liquid biopsy for cancer. Clinical Chemistry. 2015;**61**(1):112-123

[90] Douillard JY et al. Gefitinib treatment in EGFR mutated Caucasian NSCLC: Circulating-free tumor DNA as a surrogate for determination of EGFR status. Journal of Thoracic Oncology. 2014;**9**(9):1345-1353

[91] Webb S. The cancer bloodhounds. Nature Biotechnology. 2016;**34**(11):1090-1094

[92] Croce CM. Causes and consequences of microRNA dysregulation in cancer. Nature Reviews. Genetics. 2009;**10**(10):704-714

[93] Armand-Labit V, Pradines A. Circulating cell-free microRNAs as clinical cancer biomarkers. Biomolecular Concepts. 2017;**8**(2):61-81

[94] Kalluri R. The biology and function of exosomes in cancer. The Journal of Clinical Investigation. 2016;**126**(4):1208-1215

[95] Greening DW et al. A protocol for exosome isolation and characterization: Evaluation of ultracentrifugation, density-gradient separation, and immunoaffinity capture methods. Methods in Molecular Biology. 2015;**1295**:179-209

[96] Hoshino A et al. Tumour exosome integrins determine organotropic metastasis. Nature. 2015;**527**(7578):329-335

[97] Peinado H et al. Melanoma exosomes educate bone marrow progenitor cells toward a pro-metastatic phenotype through MET. Nature Medicine. 2012;**18**(6):883-891

[98] Best MG et al. RNA-Seq of tumor-educated platelets enables blood-based Pan-cancer, multiclass, and molecular pathway cancer diagnostics. Cancer Cell. 2015;**28**(5):666-676

[99] Stone RL et al. Paraneoplastic thrombocytosis in ovarian cancer. The New England Journal of Medicine. 2012;**366**(7):610-618

[100] Diem S et al. Neutrophil-to-lymphocyte ratio (NLR) and platelet-to-lymphocyte ratio (PLR) as prognostic markers in patients with non-small cell lung cancer (NSCLC) treated with nivolumab. Lung Cancer. 2017;**111**:176-181

Chapter 5

Energy Metabolism Heterogeneity-Based Molecular Biomarkers for Ovarian Cancer

Na Li, Xiaohan Zhan and Xianquan Zhan

Abstract

Energy metabolism heterogeneity is a hallmark in ovarian cancer; namely, the Warburg and reverse Warburg effects coexist in ovarian cancer. Exploration of energy metabolism heterogeneity benefits the discovery of the effective biomarkers for ovarian cancers. The integrative analysis of transcriptomics (20,115 genes in 419 ovarian cancer samples), proteomics (205 differentially expressed proteins), and mitochondrial proteomics (1198 mitochondrial differentially expressed proteins) revealed (i) the upregulations of rate-limiting enzymes PKM2 in glycolysis, IDH2 in Krebs cycle, and UQCRH in oxidative phosphorylation (OXPHOS) pathways, (ii) the upregulation of PDHB that converts pyruvate from glycolysis into acetyl-CoA in Krebs cycle, and (iii) that miRNA (hsa-miR-186-5p) and RNA-binding protein (EIF4AIII) had target sites in those key proteins in energy metabolism pathways. Furthermore, lncRNA SNHG3 interacted with miRNA (hsa-miR-186-5p) and RNA-binding protein (EIF4AIII). Those results were confirmed in the ovarian cancer cell model and tissues. It clearly concluded that lncRNA SNHG3 regulates energy metabolism through miRNA (hsa-miR-186-5p) and RNA-binding protein (EIF4AIII) to regulate the key proteins in the energy metabolism pathways. SNHG3 inhibitor might interfere with the energy metabolism to treat ovarian cancers. These findings provide more accurate understanding of molecular mechanisms of ovarian cancers and discovery of effective energy-metabolism-heterogeneity therapeutic drug for ovarian cancers.

Highlights

- Mitochondrial proteomics revealed the energy metabolism heterogeneity in ovarian cancers.
- LncRNA SNHG3 was related to ovarian cancer survival and energy metabolism with ovarian cancer TCGA analysis.
- SNHG3 was related to energy metabolism by regulating miRNAs and EIF4AIII based on GSEA and Starbase analyses.
- MiRNAs and EIF4AIII regulate the glycolysis, Krebs cycle, and OXPHOS pathways by targeting PKM, PDHB, IDH2, and UQCRH.

Keywords: ovarian cancer, iTRAQ, mitochondrial proteomics, TCGA, energy metabolism, SNHG3

1. Introduction

Ovarian cancer is a common gynecologic cancer with high mortality [1]. Despite chemotherapy, radiotherapy, surgery, and target therapy has previously been developed in ovarian cancers [2], the 5-year overall survival rate for patients who diagnosed with late stage III–IV disease is still very poor (about 30%). Because of the site of the ovaries and the certain clinical characteristics of epithelial cancers, it is a challenge to make early diagnosis [3]. Women with high-risk factors (e.g., family history, or BRCA mutations) plan for a follow-up visit with cancer antigen 125 (CA-125) monitoring and ultrasound, however, prospective validation of these physical examination or lab tests remain elusive [4]. The changes of energy metabolism are common in cancer cells, which might be potential biomarkers and therapeutics targets [5]. During the last decade, a great attention has been paid to metabolic reprogramming of cancer. However, cancer basic studies fail to reach a consistent conclusion on mitochondrial function in cancer energy metabolism [6]. The traditional view of Warburg was that cancer cells undergo aerobic glycolysis, which refers to the fermentation of glucose to lactate in the presence of oxygen as opposed to the complete oxidation of glucose, thus brought attention to the role of mitochondria in tumorigenesis [7]. A previous study found that the glycolysis enzyme PKM2 is important for cancer metabolism and tumor growth, which can improve activity and expression of PKM2 [8]. On the contrary, mitochondria were observed dysfunction, including the decreased effectiveness of Krebs cycle and electron transfer chain (ETC) complexes decoupling [9]. However, a novel 'reverse Warburg effect', was put forward in 2009 and impacted previous perceptions on cancer metabolism [10]. In this model of reverse Warburg chain, cancer cells and the cancer-associated fibroblasts (CAFs) become metabolically coupled. Interactions between cancer cells and tumor-microenvironment (TME) highly affect proliferation, energy metabolism, metastasis, and relapse of carcinoma [11]. Cancer cells secrete a large amount of ROS into microenvironment to enhance oxidative stress in CAFs. If the inflammatory reaction, autophagy, loss of stromal caveolin-1 (Cav-1), and nitric oxide synthase (NOS) are increased in CAFs, there is a good chance for progression of aerobic glycolysis [12]. Consequently, CAFs secrete plenty of energy-rich fuels to TME, including ketone bodies, lactate, pyruvate, and fatty acids. In turn, the nourishment 'feed' mitochondrial oxidative phosphorylation and ATP supplements [13]. In this process, mono-carboxylate transporters (MCTs) were highly expressed in both cancer cells and CAFs to be involved in some regulations. Immunochemistry result demonstrates that MCT4 was distributed specifically in CAFs in human breast cancers, which implicated in lactate efflux progress; while MCT1 participated in lactate uptake, and significantly upregulated specifically in kinds of cancer cells [14]. Thus evidence indicates limitations of 'the Warburg effect'. However, some studies demonstrated that aerobic glycolysis was not the dominant energy metabolism approach for many human cancer cell lines. In the past decades, studies on Warburg and reverse Warburg effects in cancers have formed a new frontier regarding additional roles of mitochondria in a cancer, and multiple functions of mitochondria have been identified in tumorigenesis [15].

High-throughput proteomics approach provides a scientific evaluation of protein expression. Functional proteomics offers more subtle clues, due to a greater attention paid to subcellular proteome research [16]. However, the subcellular proteomics of ovarian cancer mitochondrial proteins has not been elucidated. Mitochondria are the center of energy metabolism in eukaryotic cells, and also involved in other functions, such as cell signaling, cellular differentiation, cell death, and maintaining control of the cell cycle and oxidative stress regulation [17]. Those mitochondria-mediated biological processes are so closely associated with tumor relapse or metastasis. Thus, cancer therapeutics should urgently find a way to explore molecular mechanisms of mitochondrion during tumorigenesis and tumor progression [18]. Inside the cancer cell appeared structural and morphological alterations of the mitochondria, and variations of morphology and performance are presumably associated with mitochondrial differentially expressed proteins (mtDEPs) [19]. A slight increase in research on ovarian cancer has occurred in recent years, quantitative proteomic analysis of mitochondria from human ovarian cancer cells and their paclitaxel-resistant sublines proved that the chemoresistance mechanisms were partly related to the mitochondria [20]. Mitochondria similarly impart considerable flexibility for tumor cell growth and survival in otherwise harsh environments such as during nutrient depletion, hypoxia and cancer treatments, and are therefore key players in tumorigenesis [15]. The subcellular proteomics of ovarian cancer mitochondrial proteins may offer new insights into aspect of tumor development.

Regeneration of energy metabolism plays crucial roles in the pathogenesis and development of cancer since it accelerates cancer cell growth, cell cycle, proliferation and metastasis [21]. The impact of non-coding RNAs (ncRNAs) has profoundly touched the fields of human cancers, cell biology, functional genomics, and drug therapy. Long non-coding RNAs (lncRNAs) (>200 nucleotides) and microRNAs (20–24 nucleotides) have attracted much attention, which acted as key regulators in the cellular biological processes, gene expression, gene regulation, basic biological functions of eukaryotic genomes, and post-transcriptional regulation of mRNA [22]. Recent studies demonstrated that lncRNAs were widely used as biomarkers for the diagnosis and prognosis of malignant tumors [23], and some lncRNAs can even act as the new therapeutic targets [24]. More and more researchers have turned attention to the mechanism between non-coding RNAs and malignant tumors. LncRNAs affects on energy metabolism-related signaling pathways induced epigenetic regulation [25]. MicroRNAs can silence gene expression by binding to 3′ untranslated region (3'UTR) sequences in their target messenger RNAs (mRNAs), resulting in the inhibition of translation or mRNA degradation, but the interaction of lncRNAs with microRNAs can hamper this effect [26]. The present results revealed that lncRNA FOXD2-AS1 acted as a tumor promoters partly through EphB3 inhibition by directly interacting with lysine (K)-specific demethylase 1A (LSD1) and zeste homolog 2 (EZH2), which indicates that lncRNA-target gene-carcinogenesis axis for cancers does exist [27]. Here, it emphasizes important scientific associations of lncRNAs with energy metabolism in cancer cells. Increasing evidence indicates that lncRNAs play significant roles in cancer metabolism, and explore the potential mechanisms that could help elucidate regulation axis or network and provide a new direction for clinical management of different malignant phenotypes [28]. In our previous research, iTRAQ-based quantitative proteomics identified 1198 mitochondrial differentially expressed proteins (mtDEPs) between mitochondria samples isolated from human ovarian cancer and control tissues [29] and 205 differentially expressed proteins (DEPs) between human ovarian cancers and controls tissues [39]. The TCGA database includes 20,115 genes in 419 ovarian cancer samples. The conjoint analysis of 1198 mtDEPs, 205 DEPs, and 20,115 TCGA data in ovarian cancers investigated the biological pathways and molecular mechanisms of SNHG3-downstream genes-energy metabolism axis. LncRNA SNHG3 was associated with survival for ovarian cancers, and further gene set enrichment analysis proved the roles of SNHG3 in the energy metabolism through miRNAs and RNA binding protein EIF4AIII to target genes, including PKM, PDHB, IDH2, and UQCRH [29].

Figure 1 showed the experimental flow-chart of integrative analysis of 1198 mtDEPs [29], 205 DEPs [39], and 20,115 TCGA data in ovarian cancers [29] to reveal energy heterogeneity and its molecular mechanisms.

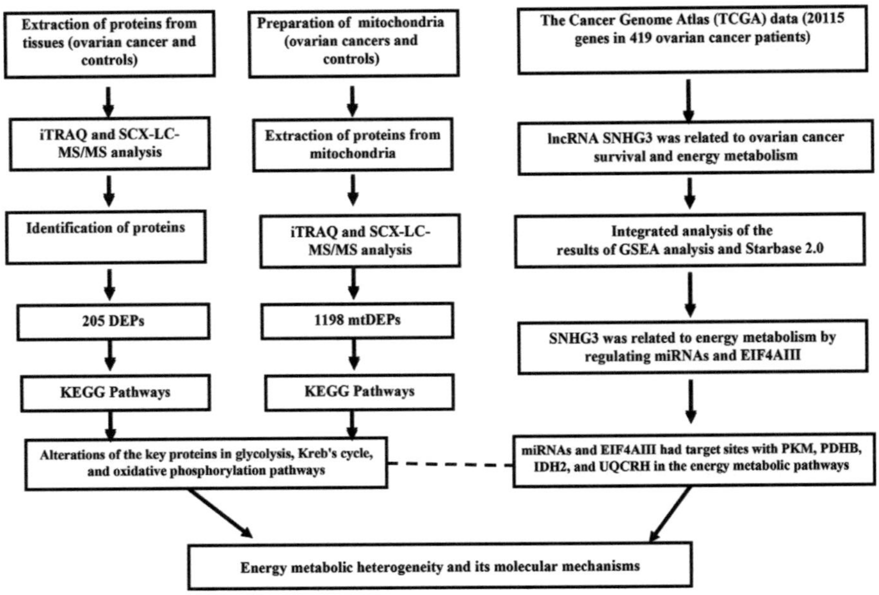

Figure 1.
The experimental flow-chart to study energy metabolic heterogeneity and its molecular mechanisms.

2. Methods

2.1 Ovarian cancer mitochondrial DEP data and bioinformatic analysis

Mitochondria were separated from 7 ovarian cancer tissues (high-degrade, poorly or moderately differentiated carcinoma cells) (cancer group) and 11 control ovaries with benign gynecologic diseases (fibroids, adenomyosis, ovary serous cystadenoma, cervical intraepithelial neoplasia, atypical hyperplasia of endometrium, and pelvic organ prolapse) (control group), respectively [29]. The separated mitochondria were validated with electron microscopy and Western blotting. The extracted proteins from the prepared mitochondrial samples were used for iTRAQ-quantitative proteomics analysis. The extracted mitochondrial proteins from ovarian cancers and controls were analyzed with 6-plex iTRAQ labeling, SCX fraction, and LC-MS/MS. MS/MS data were used to determine proteins, and the intensities of iTRAQ reporter ions were used to determine each mitochondrial DEP. The mitochondrial DEPs were further analyzed by bioinformatics including GO functional enrichment and KEGG pathway enrichment with DAVID Bioinformatics Resources 6.7.

2.2 Ovarian cancer DEP data and bioinformatic analysis

Proteins were extracted from ovarian cancer and control tissues. The extracted proteins from ovarian cancers and controls were analyzed with 6-plex iTRAQ labeling, SCX fraction, and LC-MS/MS. MS/MS data were used to determine proteins, and the intensities of iTRAQ reporter ions were used to determine each mitochondrial DEP [39]. The mitochondrial DEPs were further analyzed by bioinformatics including GO functional enrichment and KEGG pathway enrichment with DAVID Bioinformatics Resources 6.7.

2.3 TCGA data of ovarian cancer patients and bioinformatic analysis

TCGA (http://cancergenome.nih.gov/) includes 20,115 genes of 419 ovarian cancer patients, in the level of transcriptome. Those genes were classified as coding/

non-coding RNAs (mRNAs/ncRNAs) provided by the GENCODE/ENSEMBL pipeline. lncRNA genes were considered as a type of genes that exclusively produce transcripts of the 'antisense'. The lncRNA survival analysis was performed by TANRIC (http://ibl.mdanderson.org/tanric/_design/basic/index.html). The Kaplan-Meier method was used to calculate overall survival. According to median value (3.39) of SNHG3 RNA expressions, 419 ovarian cancer patients were divided into SNHG3 high (>3.39; n = 210) vs. low (<3.39; n = 209) expression groups. TCGA data of two groups were analyzed with GSEA enrichment analysis. Moreover, the lncRNA expressions from Cancer Cell Line Encyclopedia (https://portals.broadinstitute.org/ccle), and chemosensitivity of tamoxifen from Genomics of Drug Sensitivity in Cancer (http://www.cancerrxgene.org/) were obtained for ovarian cancer cell lines. GraphPad Prism v6.0 (GraphPad Software, San Diego, CA, USA) was used to construct histogram.

2.4 Integrative analysis of mitochondrial DEPs, tissue DEPs, and TCGA data with bioinformatics

The integrated miRNA-lncRNA SNHG3, miRNA-target gene, RNA binding protein-lncRNA SNHG3, RNA binding protein-mRNA, and protein-protein signatures were identified. STRING 10.0 was used predict interactions of chemicals and proteins. Chemicals were linked to other chemicals and proteins by evidence derived from experiments, databases and literature (http://string-db.org/cgi/input.pl). The large-scale CLIP-Seq data by starBasev 2.0 (http://starbase.sysu.edu.cn/mirCircRNA.php) was used to construct SNHG3-miRNA, protein-miRNA, SNHG3-RNA binding protein, mRNA-RNA binding protein, and mRNA-microRNA-lncRNA interaction networks. The mitochondrial DEPs in ovarian cancers were input into STRING for protein–protein interaction analysis. Network visualizations were performed with Cytoscape 3.4.0 (http://www.cytoscape.org/). The binding sites of 3′UTR region of targeted genes were predicted with three publicly available databases (TargetScan, NCBI, and RNAhybrid), sequences of microRNA (>hsa-miR-186-5p MIMAT0000456 CAAAGAAUUCUCCUUUUGGGCU) and PDHB 3′UTR region. MicroRNA binding sites with PDHB were predicted with RNAhybrid database.

2.5 Experimental validation in cell models

Three ovarian cancer cell lines (TOV-21G, SK-OV3, and OVCAR-3), and one normal control cell line (IOSE80) from Keibai Academy of Science (Nanjing, China) were used. RPMI-1640 medium were used to culture TOV-21G and OVCAR-3 cells in 5% CO_2 atmosphere at 37°C. DMEM medium (Corning, NY, USA) were used to culture IOSE80 and SK-OV3 in 5% CO_2 atmosphere at 37°C, with supplementation of 10% fetal bovine serum (FBS, GIBCO, South America, NY, USA). (i) Transient transfection was performed with Lipofectamine 3000 reagents according to the manufacturer's instructions (Invitrogen, USA). SK-OV3, OVCAR-3, and TOV-21G were seeded in 6-well plates at 30–50% density. Cells were collected at 24–48 h after transfection, for next-step experiments. (ii) RNA extraction and quantitative real-time PCR (qRT-PCR) analyses. TRizol® Reagent (Invitrogen, CA, USA) was used to extract total RNAs. total RNAs were reversely transcribed into cDNAs and then used to perform qRT-PCR analysis to detect SNHG3 and its target genes, with β-actin as an internal control. (iii) 1D-SDS-PAGE and Western blotting was used to detect PKM, PFKM, PDHB, IDH2, CS, OGDHL, and UQCRH against the corresponding antibodies, with β-actin as internal control. (iv) Data were expressed as the mean ± SD of triplicates. Each experiment was repeated at least three times. In all cases, $P < 0.05$ was considered as statistical significance.

3. Results and discussion

3.1 The changes of key proteins in the energy metabolism signaling pathways

The iTRAQ-based quantitative proteomics identified 1198 DEPs between mitochondria samples isolated from ovarian cancer and control tissues [29]. The statistically significant KEGG pathways were mined with DAVID Bioinformatics Resources from those mitochondrial DEPs between EOCs and controls, among which those DEPs were significantly enriched in the processes of Krebs cycle, and oxidative phosphorylation (OXPHOS) pathways. The key proteins (PDHB, IDH2, and UQCRH) were associated with aerobic oxidation to supply in the Krebs cycle, and oxidative phosphorylation was upregulated (**Figure 2**). Interestingly, those results were coincided with the reverse Warburg effect proposed in 2009 [10].

The iTRAQ-based quantitative proteomics identified 205 DEPs between ovarian cancer and control tissues [13], which revealed the upregulation of the key enzyme PKM2 in glycolysis pathway I in ovarian cancers. It was coincided with the Otto Warburg effect proposed in 1956 [30]. Warburg discovered that cancer cells tend to produce ATP by aerobic glycolysis, even though it is a less efficient pathway contrasted with OXPHOS. This popular system, called 'Warburg effect', has been the dominant mechanism of tumors for energy generations, while its relationship with tumorigenesis remains still unclear.

The research of the 'Warburg effect' mechanism of a cancer cell has never interrupted at home and abroad. PKM2, a splice isoform of the pyruvate kinase, serves as a major metabolic reprogramming regulator with an adjustable activity subjected to numerous allosteric effectors and post-translational modifications [31]. One observed that PKM2 modification was associated with an enhanced glucose consumption, level of lipid and DNA synthesis, and lactate productions, indicating

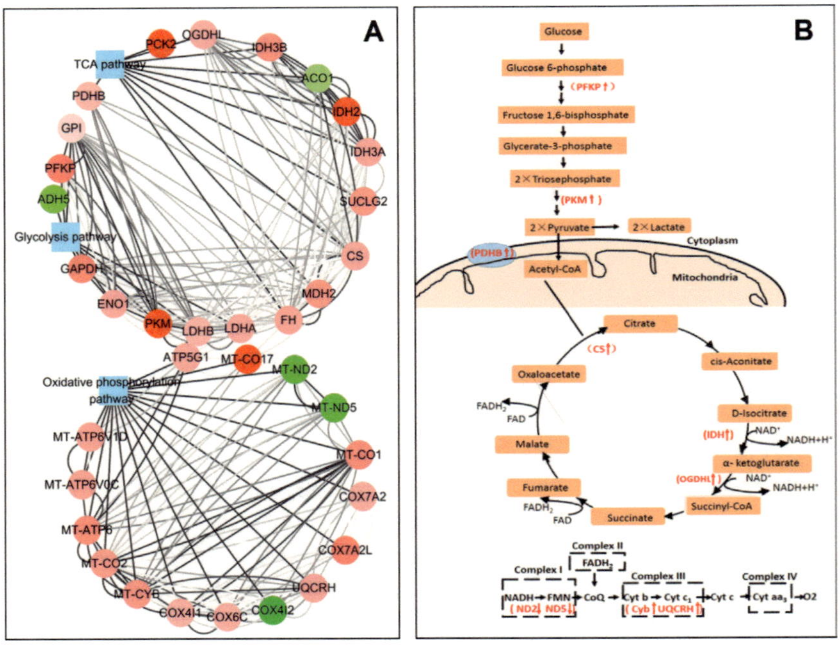

Figure 2.
The changes of key proteins in oxidative phosphorylation, Krebs cycle, and glycolysis pathways. A. The network derived from identified genes and pathways. The square box was pathways. The circular box was DEPs. The edge was the gene–gene and gene-pathway interactions. The node color from green to red was the gene expression level from low to high. The edge color was the correlation level from low to high (gray = 0.70–0.90, black = 0.90–1.00). B. Oxidative phosphorylation, Krebs cycle, and glycolysis pathways were altered in ovarian cancers. Reproduced from Li et al. [29], with permission from Elsevier, copyright 2018.

that PKM2 transformation promotes the Warburg effect [32]. Then, a novel series of inhibitors were developed to anti-Warburg-effect drugs for cancer treatment. For example, erastin-like anti-Warburg agents prevent mitochondrial depolarization induced by free tubulin and decrease lactate formation in cancer cells [33]. However, Warburg effect also has some limitations, because it completely ignored these facts that cancer cells had a great interaction with tumor microenvironment.

In the 2009, a new model for understanding the Warburg effect was proposed in tumor energy metabolism. The hypothesis is that cancer cells induce the aerobic glycolysis in neighboring stromal fibroblasts. These cancer-associated fibroblasts (CAFs) secrete energy-rich substances, including lactate and pyruvate, to tumor microenvironment. These energy-rich metabolites were eaten up by adjacent cancer cell and used by mitochondrial TCA cycle, resulting in a higher energy producing capacity. It termed this new idea as the "reverse Warburg effect" [10]. Taken all together, the reverse Warburg effect is a new energy metabolic pattern identified between cancer cells and CAFs, but this novel pattern does not deny Warburg effect status and still cannot replace it. Actually, the reverse Warburg effect extends energy metabolism content, which explained the nature of the heterogeneity and plasticity of cancer metabolism [34]. Although it's validated that the 'reverse Warburg effect' can be initiated by oxidative stress in two compartment metabolic coupling and change of cellular electromagnetic filed, detailed mechanisms remain still unclear.

In order to verify the above of views, each EOC cell line (SK-OV3, TOV-21G, and OVCAR-3) showed high expression of energy metabolism-related genes relative to control cells IOSE80 by qRTPCR, such as PKM, PDHB, IDH3A, IDH3B, ND5, ND2, and CYB in EOC cell lines relative to IOSE80 (**Figure 3**).

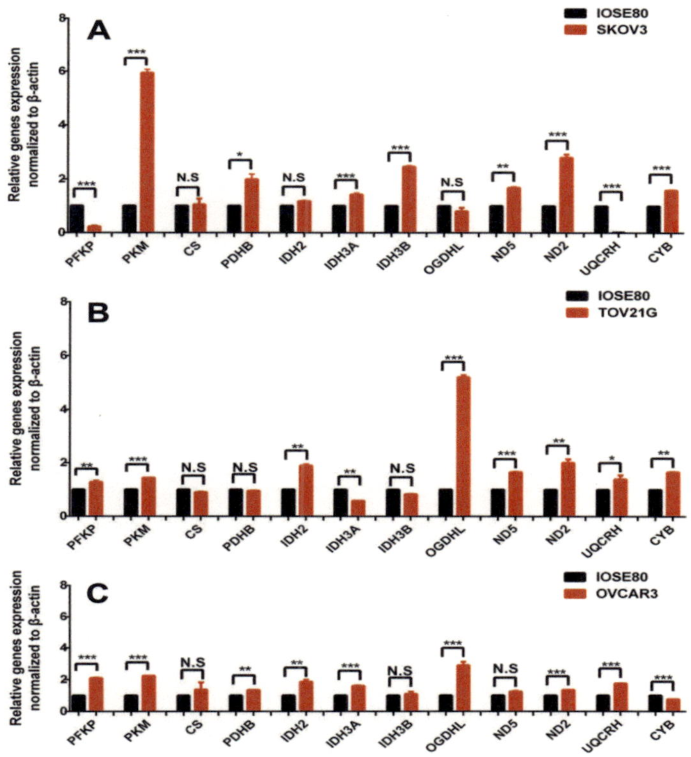

Figure 3.
The gene expression changes of key proteins in glycolysis, Krebs cycle, and oxidative phosphorylation pathways confirmed by qRT-PCR analysis in ovarian cancer cells. $^{*}p < 0.05$; $^{**}p < 0.01$; $^{***}p < 0.001$. *N.S., non-significance. Reproduced from Li et al. [29], with permission from Elsevier, copyright 2018.*

3.2 SNHG3 was significantly related to EOC survival through the key molecules in the energy metabolism pathways by their RNA-binding proteins or miRNA

Among identified 1198 mitochondrial DEPs and 205 tissue DEPs, PFKM, PKM, PDHB, CS, IDH2, IDH3A, IDH3B, OGDHL, ND2, ND5, CYB, and UQCRH were enriched in glycolysis, Krebs cycle, and oxidative phosphorylation pathways. Six RNA-binding proteins (EIF4AIII, IGF2BP2, C22ORF28, UPF1, SFRS1, and EWSR1) were the iTRAQ-identified proteins in ovarian cancers (**Figure 4A**) based on Starbase v2.0 database. However, only EIF4AIII was associated with energy metabolic pathway, when did protein-protein network by STRING 10.0 software (**Figure 4B**). Furthermore, overlapping analysis between survival-related lncRNAs of EOC and lncRNAs binding with EIF4AIII obtained 16 lncRNAs (LINC00517, SNHG3, LBX2-AS1, ZNRF3-AS1, LINC00565, AL109767.1, WWTR1-AS1, HCG15, LEMD1-AS1, PDCD4-AS1, KIF9-AS1, SOS1-IT1, STARD13-IT1, PLCH1-AS1, ZNF674-AS1, and HOXC-AS3) existed those two groups. Among those 16 overlapped lncRNAs, only lncRNA SNHG3 was associated with energy metabolic pathways by GSEA analysis (**Figure 4C** and **D** and **Figure 4A** and **B**). The expression levels of lncRNA SNHG3 in different ovarian cancer cell lines indicated that poorly differentiated cell lines existed high SNHG3 expression, such as TYK-nu ovarian cancer cell line (**Figure 4E**). Additionally, q-PCR data demonstrated that SNHG3 was upregulated in SKOV3, TOV21G, and OVCAR3 relative to control cell line (IOSE80) (**Figure 4F**).

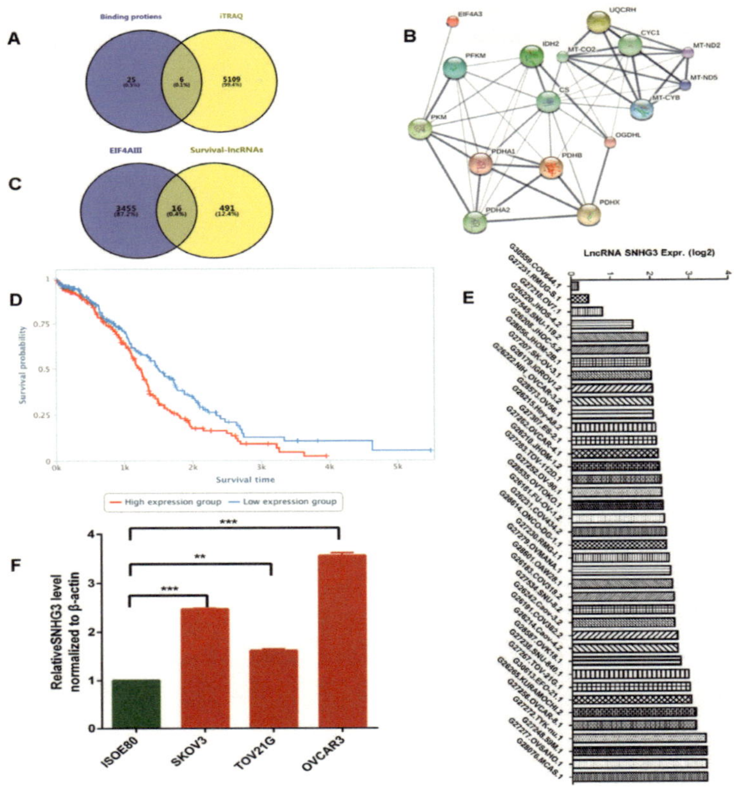

Figure 4.
*SNHG3 was significantly related to ovarian cancer survival through the key molecules in the energy metabolism pathways by their RNA-binding proteins or miRNA. A. Overlapping analysis of identified proteins and RNA binding proteins. B. Target DEPs-based protein-protein interaction network (STRING 10.0). C. Overlapping analysis of EIF4AIII-binding lncRNAs and lncRNAs involved in ovarian cancer survival. D. Kaplan–Meier survival analysis based on ovarian cancer SNHG3. E and F. SNHG3 expressions in ovarian cell lines. **$p < 0.01$; ***$p < 0.001$. Reproduced from Li et al. [29], with permission from Elsevier, copyright 2018.*

One should also notice that gene sets were also significantly enriched in pathway of tamoxifen response (**Figure 5**), which indicated SNHG3 was associated with drug sensitivity or multidrug resistance. The comprehensive evaluation on SNHG3 may lead to ways to improve drug sensitivity of tamoxifen in EOCs.

Gene sets enrichment analysis showed that mRNA metabolism and 3′UTR-mediated translational regulation (**Figure 6**). Overlap analysis of RNAs-RNAs interaction networks showed that SNHG3 may regulated PDHB through binding hsa-miR-186-5p or hsa-miR-590-3p (**Figure 7A**), especially, hsa-miR-186-5p obtained high stringency to target PDHB with Starbase 2.0 analysis. Meanwhile, two binding sites were predicted between putative hsa-miR-186-5p and PDHB 3’UTR with RNAhybrid database (**Figure 7B** and **C**). Here, it can be forecasted boldly that SNHG3 might regulate the EOC energy metabolism by binding EIF4AIII and hsa-miR-186-5p, functioned as efficient sponges to regulate energy metabolism pathways though mitochondrial key molecules (**Figures 7D** and **8A** and **B**).

To further verify that SNHG3 can lead to the carcinogenesis in vivo, SKOV3 cells were transfected with either si-SNHG3 or a si-RNA negative control. Target genes, including PFKM, PKM, PDHB, CS, IDH2, IDH3A, IDH3B, OGDHL, ND5, ND2, CYB and UQCRH turned significant decrease expression (**Figure 9**). The results were further validated to a reasonable degree by Western blot (**Figure 10**).

Non coding RNAs, as one of epigenetic regulation form, play an important role in activation and suppression in a tumor by altering cell energy metabolism or biological behaviors [35]. However, lncRNAs have been identified and reported to be related to many kinds of carcinomas, little is known about lncRNAs whole molecular mechanisms in tumor energy metabolism. Recently, discovery of novel biomarkers focuses on ncRNAs, such as miR-125a, MALAT1, let-7a, miR-196a, HOXA11-AS, and lncRNA FAL1 [36]. Some biomarkers have been verified consistency in both tissues and serum, which improved clinical application value to use in early diagnosis or monitoring patient prognosis [37]. A number of studies have shown that lncRNAs can play an important role in tumorigenesis and progression through a variety of mechanisms, such as binding transcription factor, acting as miRNA sponge, ceRNA (competing endogenous RNAs) [38].

Therefore, lncRNA as an effective screening and their potential mechanisms in tumor energy metabolism would be rather influential in EOCs.

3.3 Potential therapeutic targets in metabolic symbiosis

Tumor tissues were made up by parenchymal cells and stromal elements. Parenchymal cells probably showed metabolism heterogeneity. So some cancer cells were high glycolytic cancer cells consisting with “Warburg effect”, while

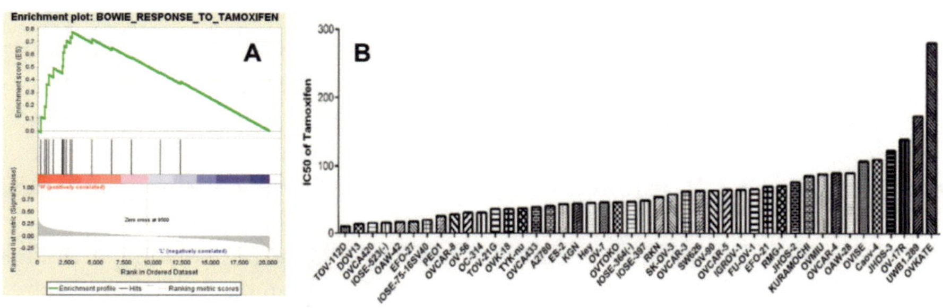

Figure 5.
The results of SNHG3 by GSEA analysis revealed tamoxifen pathway. A. Genes were enriched in response to tamoxifen. B. IC50 of tamoxifen in EOC cell lines. Reproduced from Li et al. [29], with permission from Elsevier, copyright 2018.

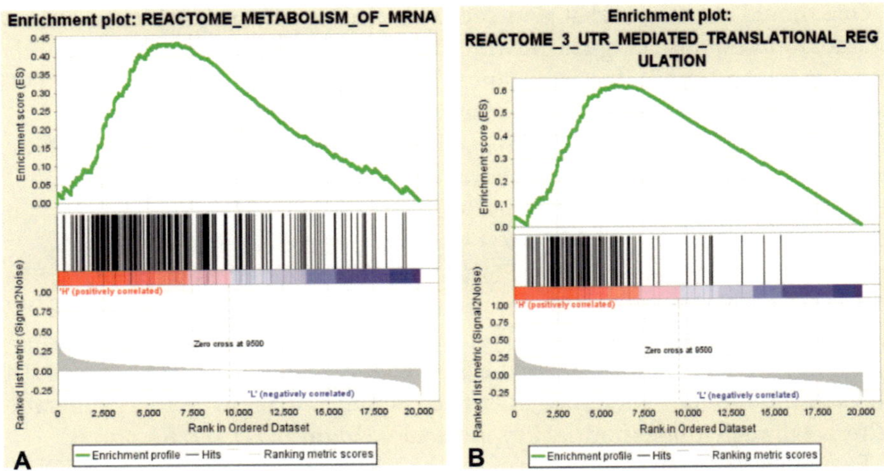

Figure 6.
The results of SNHG3 by GSEA analysis revealed mRNA metabolism and reactome 3′UTR-mediated translational regulation. A. Genes were enriched in metabolism of mRNAs. B. Genes were mainly enriched in reactome 3′UTR-mediated translational regulation. Reproduced from Li et al. [29], with permission from Elsevier, copyright 2018.

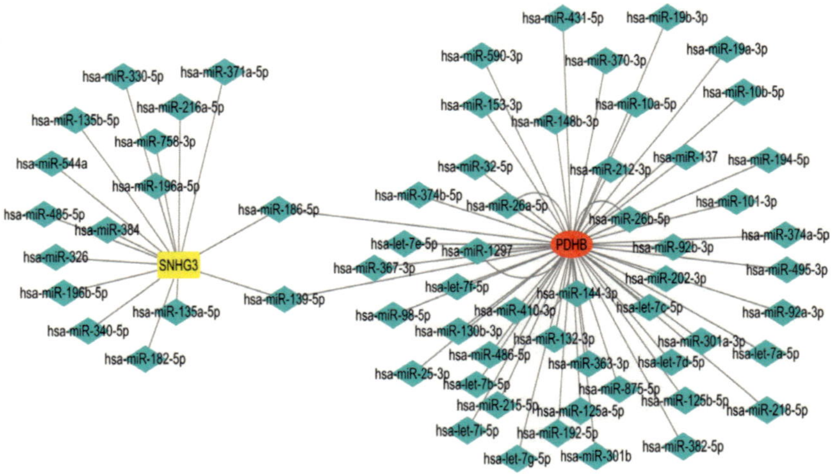

Figure 7.
Overlapping analysis of the microRNA/mRNA and microRNA/lncRNAs interactive network. Reproduced from Li et al. [29], with permission from Elsevier, copyright 2018.

other cancer cells were oxidative cancer cells consisting with "reverse Warburg effect". Cancer cells and stroma cells (especially CAFs) have metabolic symbiosis, so cancer cells induce oxidative stress of CAFs by secreting ROS to enhance aerobic glycolysis of CAFs. In turn, CAFs produced lots of nourishment to be 'eaten' up by cancer cells for producing ATP through Krebs cycle and oxidative phosphorylation [13].

MCT-1 and MCT-4 were overexpressed in EOC cells by qRT-PCR experiments, including SKOV3, TOV21G and OVCAR3 (**Figure 11**). Even though tumors were characterized by metabolic heterogeneity, MCT-1 and MCT-4 were just like lactate shuttle between cancer cells and stroma cells. The nanomaterial-siRNAs of SNHG3 might be promising for EOC patients to block the abnormal energy metabolism (**Figure 12**).

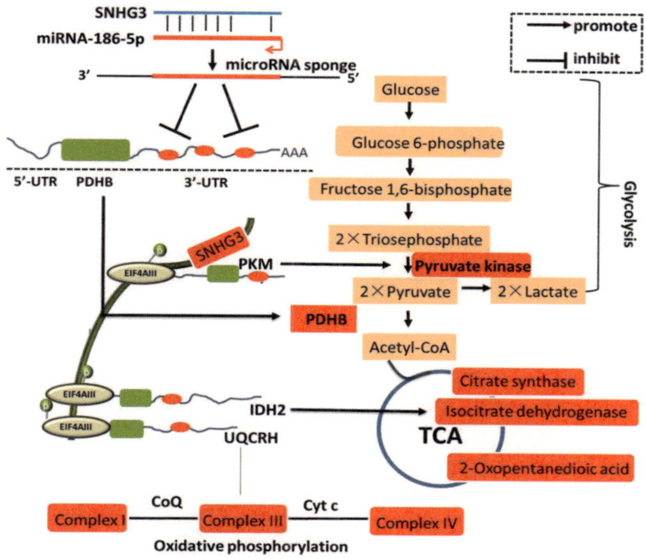

Figure 8.
Schematic model of the potential signaling mechanisms between SNHG3 and energy metabolism in the EOC regulation. Reproduced from Li et al. [29], with permission from Elsevier, copyright 2018.

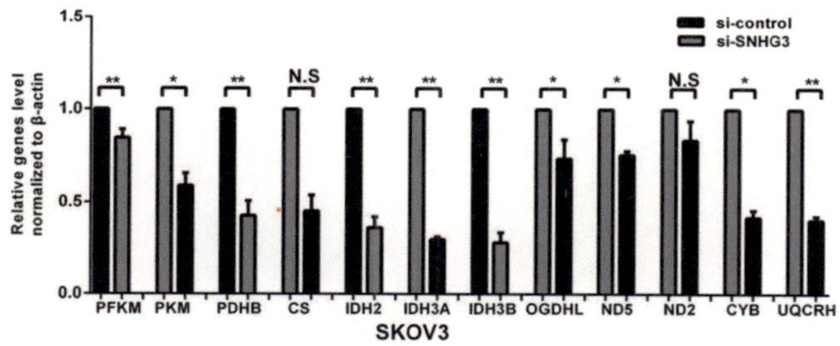

Figure 9.
The mRNA expression levels of target genes of SNHG3 in EOC cells were determined by qRT-PCR. $^*p < 0.05$, $^{**}p < 0.01$. *N.S., non-significance. Reproduced from Li et al. [29], with permission from Elsevier, copyright 2018.*

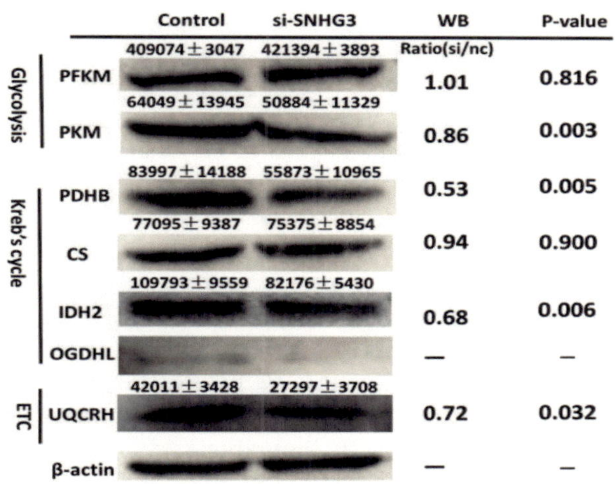

Figure 10.
The protein expression levels of target genes of SNHG3 in EOC cells were determined by Western blot. Reproduced from Li et al. [29], with permission from Elsevier, copyright 2018.

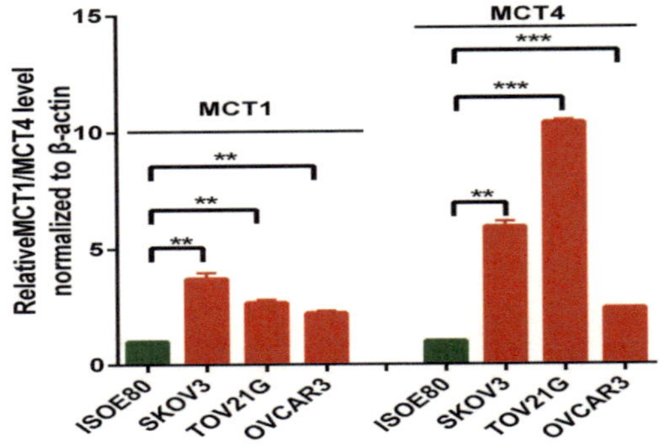

Figure 11.
*The expression levels of MCT-1 and MCT-4 in EOC cells were determined by qRT-PCR. **p < 0.01, ***p < 0.001. N.S., non-significance. Reproduced from Li et al. [29], with permission from Elsevier, copyright 2018.*

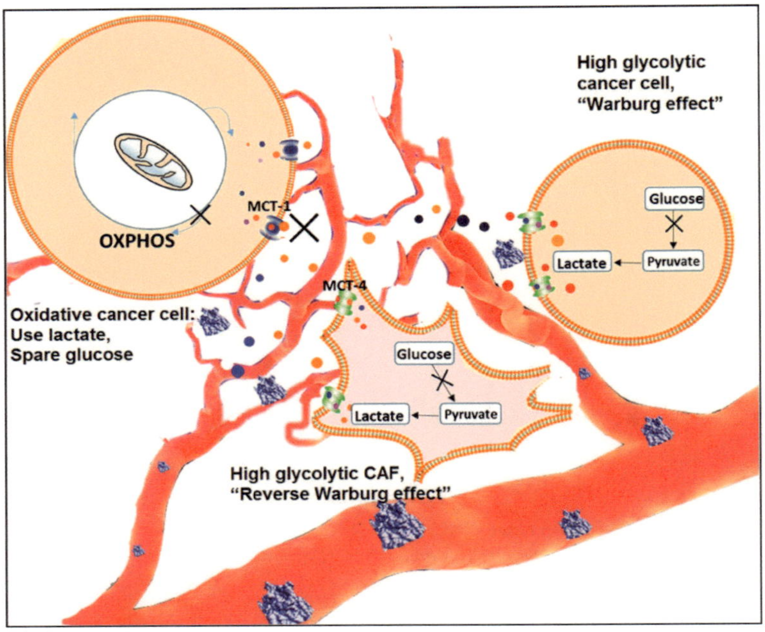

Figure 12.
Energy metabolic heterogeneity-based potential therapeutic targets model. Parenchymal cells demonstrated energy metabolic heterogeneity. Some cancer cells showed the "Warburg effect" with highly glycolytic functions, and other cancer cells showed the "reverse Warburg effect" with oxidative cancer cells. The metabolic symbiosis existed between tumor cells and CAFs through MCTs. The RNA interference sequence of SNHG3 might be effective. Modified from Li et al. [29], with permission from Elsevier, copyright 2018.

4. Conclusions

The identified 1198 mitochondrial DEPs, 205 tissue DEPs, and TCGA data in ovarian cancers provide new insights into human ovarian cancers, particularly the energy metabolism heterogeneity that 'Warburg effect' and 'reverse Warburg effect' were coexisted in ovarian cancer tissues. It emphasizes the important scientific merit in identity of new useful biomarkers within EOC energy metabolism heterogeneity system for the diagnosis and prognosis of ovarian cancer, and discovery of some potential therapeutic targets in energy metabolic interactions. Moreover, SNHG3 was related to energy metabolism through regulating hsa-miR-186-5p or RNA binding protein EIF4AIII, and those two molecules had target sites with key proteins in TCA cycle and oxidative phosphorylation pathways (PDHB, IDH2, and UQCRH). Therefore, energy metabolism-based target treatments might be very promising for ovarian cancer patients to block both 'Warburg effect' and 'reverse Warburg effect'.

Acknowledgements

The authors acknowledge the financial supports from the Xiangya Hospital Funds for Talent Introduction (to X.Z.), the Hunan Provincial "Hundred Talent Plan" program (to X.Z.), the National Natural Science Foundation of China (Grant nos. 81572278 and 81272798 to X.Z.), China "863" Plan Project (Grant No. 2014AA020610-1 to X.Z.), and the Hunan Provincial Natural Science Foundation of China (Grant No. 14JJ7008 to X.Z.).

Conflict of interest

We declare that we have no financial and personal relationships with other people or organizations.

Author's contributions

N.L. prepared figures, and wrote manuscript draft of the book chapter. X.H.Z. participated in writing and revising manuscript. X.Z. conceived the concept, designed the book chapter, prepared figures, wrote and critically revised the book chapter, coordinated and was responsible for the correspondence work and financial support.

Acronyms and abbreviations

CAFs	cancer-associated fibroblasts
DEPs	differentially expressed proteins
ETC	electron transfer chain
iTRAQ	isobaric target for relative and absolute quantification
LC	liquid chromatography
lncRNAs	long non-coding RNAs
MCTs	mono-carboxylate transporters
mRNAs	messenger RNAs
MS/MS	tandem mass spectrometry
mtDEPs	mitochondrial differentially expressed proteins
ncRNAs	non-coding RNAs
NOS	nitric oxide synthase
OXPHOS	oxidative phosphorylation
SCX	strong cation exchange
TME	tumor-microenvironment

Author details

Na Li[1,2], Xiaohan Zhan[1,2] and Xianquan Zhan[1,2*]

1 Key Laboratory of Cancer Proteomics of Chinese Ministry of Health, Xiangya Hospital, Central South University, Changsha, China

2 State Local Joint Engineering Laboratory for Anticancer Drugs, Xiangya Hospital, Central South University, Changsha, China

*Address all correspondence to: yjzhan2011@gmail.com

References

[1] Torre LA, Trabert B, DeSantis CE. Ovarian cancer statistics. CA: A Cancer Journal for Clinicians. 2018;**68**:284-296. DOI: 10.3322/caac.21456

[2] Kreienbring K, Franz A, Richter R, Dragun D, Heidecke H, Muller D, et al. The role of PAR1 autoantibodies in patients with primary epithelial ovarian cancer. Anticancer Research. 2018;**38**:3619-3625. DOI: 10.21873/anticanres.12636

[3] Luo L, Dong LY, Yan QG, Cao SJ, Wen XT, Huang Y, et al. Research progress in applying proteomics technology to explore early diagnosis biomarkers of breast cancer, lung cancer and ovarian cancer. Asian Pacific Journal of Cancer Prevention. 2014;**15**:8529-8538

[4] Urban RR, Pappas TC, Bullock RG, Munroe DG, Bonato V, Agnew K, et al. Combined symptom index and second-generation multivariate biomarker test for prediction of ovarian cancer in patients with an adnexal mass. Gynecologic Oncology. 2018. DOI: 10.1016/j.ygyno.2018.06.004

[5] Antonosante A, d'Angelo M, Castelli V, Catanesi M, Iannotta D, Giordano A, et al. The involvement of PPARs in the peculiar energetic metabolism of tumor cells. International Journal of Molecular Sciences. 2018;**19**(7). pii: E1907. DOI: 10.3390/ijms19071907

[6] Sebastian C. Tracking down the origin of cancer: Metabolic reprogramming as a driver of stemness and tumorigenesis. Critical Reviews in Oncogenesis. 2014;**19**:363-382

[7] Lu J, Tan M, Cai Q. The Warburg effect in tumor progression: Mitochondrial oxidative metabolism as an anti-metastasis mechanism. Cancer Letters. 2015;**356**:156-164. DOI: 10.1016/j.canlet.2014.04.001

[8] Christofk HR, Vander Heiden MG, Harris MH, Ramanathan A, Gerszten RE, Wei R, et al. The M2 splice isoform of pyruvate kinase is important for cancer metabolism and tumour growth. Nature. 2008;**452**:230-233. DOI: 10.1038/nature06734

[9] Birsoy K, Wang T, Chen WW, Freinkman E, Abu-Remaileh M, Sabatini DM. An essential role of the mitochondrial electron transport chain in cell proliferation is to enable aspartate synthesis. Cell. 2015;**162**:540-551. DOI: 10.1016/j.cell.2015.07.016

[10] Pavlides S, Whitaker-Menezes D, Castello-Cros R, Flomenberg N, Witkiewicz AK, Frank PG, et al. The reverse Warburg effect: Aerobic glycolysis in cancer associated fibroblasts and the tumor stroma. Cell Cycle. 2009;**8**:3984-4001. DOI: 10.4161/cc.8.23.10238

[11] Catteau X, Simon P, Noel JC. Stromal expression of matrix metalloproteinase 2 in cancer-associated fibroblasts is strongly related to human epidermal growth factor receptor 2 status in invasive breast carcinoma. Molecular and Clinical Oncology. 2016;**4**:375-378. DOI: 10.3892/mco.2015.721

[12] Lee M, Yoon JH. Metabolic interplay between glycolysis and mitochondrial oxidation: The reverse Warburg effect and its therapeutic implication. World Journal of Biological Chemistry. 2015;**6**:148-161. DOI: 10.4331/wjbc.v6.i3.148

[13] Bar-Or D, Carrick M, Tanner A 2nd, Lieser MJ, Rael LT, Brody E. Overcoming the Warburg effect: Is it the key to survival in sepsis? Journal of Critical Care. 2018;**43**:197-201. DOI: 10.1016/j.jcrc.2017.09.012

[14] Whitaker-Menezes D, Martinez-Outschoorn UE, Lin Z, Ertel A,

Flomenberg N, Witkiewicz AK, et al. Evidence for a stromal-epithelial "lactate shuttle" in human tumors: MCT4 is a marker of oxidative stress in cancer-associated fibroblasts. Cell Cycle. 2011;**10**:1772-1783. DOI: 10.4161/cc.10.11.15659

[15] Vyas S, Zaganjor E, Haigis MC. Mitochondria and cancer. Cell. 2016;**166**:555-566. DOI: 10.1016/j.cell.2016.07.002

[16] Williams EG, Wu Y, Ryu D, Kim JY, Lan J, Hasan M, et al. Quantifying and localizing the mitochondrial proteome across five tissues in a mouse population. 2018;**17**:1766-1777. DOI: 10.1074/mcp.RA118.000554

[17] Sassano ML, van Vliet AR, Agostinis P. Mitochondria-associated membranes as networking platforms and regulators of cancer cell fate. Frontiers in Oncology. 2017;**7**:174. DOI: 10.3389/fonc.2017.00174

[18] Strickertsson JAB, Desler C, Rasmussen LJ. Bacterial infection increases risk of carcinogenesis by targeting mitochondria. Seminars in Cancer Biology. 2017;**47**:95-100. DOI: 10.1016/j.semcancer.2017.07.003

[19] Mintz HA, Yawn DH, Safer B, Bresnick E, Liebelt AG, Blailock ZR, et al. Morphological and biochemical studies of isolated mitochondria from fetal, neonatal, and adult liver and from neoplastic tissues. The Journal of Cell Biology. 1967;**34**:513-523

[20] Chen M, Huang H, He H, Ying W, Liu X, Dai Z, et al. Quantitative proteomic analysis of mitochondria from human ovarian cancer cells and their paclitaxel-resistant sublines. Cancer Science. 2015;**106**:1075-1083. DOI: 10.1111/cas.12710

[21] Scalise M, Pochini L, Galluccio M, Console L, Indiveri C. Glutamine transport and mitochondrial metabolism in cancer cell growth. Frontiers in Oncology. 2017;7:306. DOI: 10.3389/fonc.2017.00306

[22] Huarte M. The emerging role of lncRNAs in cancer. Nature Medicine. 2015;**21**:1253-1261. DOI: 10.1038/nm.3981

[23] Mou Y, Wang D, Xing R, Nie H, Mou Y, Zhang Y, et al. Identification of long noncoding RNAs biomarkers in patients with hepatitis B virus-associated hepatocellular carcinoma. Cancer Biomarkers. 2018. DOI: 10.3233/CBM-181424

[24] Rodriguez Bautista R, Ortega GA. Long non-coding RNAs: Implications in targeted diagnoses, prognosis, and improved therapeutic strategies in human non- and triple-negative breast cancer. 2018;**10**:88. DOI: 10.1186/s13148-018-0514-z

[25] Chen B, Li Y, He Y, Xue C, Xu F. The emerging roles of long non-coding RNA in gallbladder cancer tumorigenesis. Cancer Biomarkers. 2018;**22**:359-366. DOI: 10.3233/cbm-170979

[26] Chen Z, Wu J, Huang W, Peng J, Ye J, Yang L, et al. Long non-coding RNA RP11-789C1.1 suppresses epithelial to mesenchymal transition in gastric cancer through the RP11-789C1.1/MiR-5003/E-cadherin axis. Cellular Physiology and Biochemistry. 2018;**47**:2432-2444. DOI: 10.1159/000491617

[27] Xu TP, Wang WY, Ma P, Shuai Y, Zhao K, Wang YF, et al. Upregulation of the long noncoding RNA FOXD2-AS1 promotes carcinogenesis by epigenetically silencing EphB3 through EZH2 and LSD1, and predicts poor prognosis in gastric cancer. Oncogene. 2018. DOI: 10.1038/s41388-018-0308-y

[28] Hua Q, Mi B, Huang G. The emerging co-regulatory role of long noncoding RNAs in

epithelial-mesenchymal transition and the Warburg effect in aggressive tumors. Critical Reviews in Oncology/Hematology. 2018;**126**:112-120. DOI: 10.1016/j.critrevonc.2018.03.028

[29] Li N, Zhan X, Zhan X. The lncRNA SNHG3 regulates energy metabolism of ovarian cancer by an analysis of mitochondrial proteomes. Gynecologic Oncology. 2018. DOI: 10.1016/j.ygyno.2018.06.013

[30] Warburg O. On the origin of cancer cells. Science. 1956;**123**:309-314

[31] Kobierzycki C, Piotrowska A, Latkowski K, Zabel M, Nowak-Markwitz E, Spaczynski M, et al. Correlation of pyruvate kinase M2 expression with clinicopathological data in ovarian cancer. Anticancer Research. 2018;**38**:295-300. DOI: 10.21873/anticanres.12221

[32] Wang Y, Liu J, Jin X, Zhang D, Li D, Hao F, et al. O-GlcNAcylation destabilizes the active tetrameric PKM2 to promote the Warburg effect. Proceedings of the National Academy of Sciences of the United States of America. 2017;**114**:13732-13737. DOI: 10.1073/pnas.1704145115

[33] DeHart DN, Lemasters JJ, Maldonado EN. Erastin-like anti-Warburg agents prevent mitochondrial depolarization induced by free tubulin and decrease lactate formation in cancer cells. SLAS Discovery. 2018;**23**:23-33. DOI: 10.1177/2472555217731556

[34] Yoshida GJ. Metabolic reprogramming: The emerging concept and associated therapeutic strategies. Journal of Experimental & Clinical Cancer Research. 2015;**34**:111. DOI: 10.1186/s13046-015-0221-y

[35] Liu P, Tang H, Wu J, Qiu X, Kong Y, Zhang L, et al. Linc01638 promotes tumorigenesis in HER2+ breast cancer. Current Cancer Drug Targets. 2018. DOI: 10.2174/1568009618666180709163718

[36] Adamopoulos PG, Tsiakanikas P, Scorilas A. Kallikrein-related peptidases and associated microRNAs as promising prognostic biomarkers in gastrointestinal malignancies. Biological Chemistry. 2017;**399**:821-836. DOI: 10.1515/hsz-2017-0342

[37] Wang Y, Zhao Z, Zhang S, Li Z, Li D, Yang S, et al. LncRNA FAL1 is a negative prognostic biomarker and exhibits pro-oncogenic function in osteosarcoma. 2018. DOI: 10.1002/jcb.27074

[38] Tian L, He Y, Zhang H, Wu Z, Li D, Zheng C. Comprehensive analysis of differentially expressed profiles of lncRNAs and mRNAs reveals ceRNA networks in the transformation of diffuse large B-cell lymphoma. Oncology Letters. 2018;**16**:882-890. DOI: 10.3892/ol.2018.8722

[39] Wang LN, Tong SW, Hu HD, et al. Quantitative proteome analysis of ovarian cancer tissues using a iTRAQ approach. Journal of Cellular Biochemistry. 2012;**113**(12):3762-3772

Chapter 6

A Novel P53/POMC/Gas/SASH1 Autoregulatory Feedback Loop and Pathologic Hyperpigmentation

Ding'an Zhou, Jiawei Zeng, Xing Zeng, Yadong Li, Zhixiong Wu, Xin Wan, Pingshen Hu and Xiaodong Su

Abstract

P53-regulated proteins in transcriptional level are associated with many signal transduction pathways and p53 plays a pivotal role in a number of positive and negative autoregulatory feedback loops. Although POMC/α-MSH productions induced by ultraviolet (UV) are directly mediated by p53, p53 is related to UV-independent pathological pigmentation. In the process of identifying the causative gene of dyschromatosis universalis hereditaria (DUH), three mutations encoding amino acid substitutions were found in the gene SAM and SH3 domain containing 1 (SASH1). SASH1 was identified to interact with guanine nucleotide-binding protein subunit-alpha isoforms short (Gαs). However, for about 90 years, the pathological gene and the pathological mechanism of DUH are unclear. Our study indicates that SASH1 is physiologically medicated by p53 upon UV stimulation and a reciprocal SASH-p53 inducement is existed physiologically and pathophysiologically. A novel p53/POMC/α-MSH/Gαs/SASH1 signal cascade regulates SASH1 to foster melanogenesis. SASH1 mutations control a novel p53/POMC/Gαs/SASH1 autoregulatory positive feedback loop to promote pathological hyperpigmentation phenotype in DUH-affected individuals. Our work illustrates a novel p53/POMC/Gαs/SASH1 autoregulatory positive feedback loop that is mediated by SASH1 mutations to foster pathological hyperpigmentation phenotype.

Keywords: SASH1 substitution mutations, p53, autoregulatory feedback loop, DUH, pathological hyperpigmentation phenotype

1. Introduction

The skin pigmentation is formed by the synthesis of melanin in the melanocytes. Melanocyte is a kind of epithelial cells mainly locating basal cell layers of epidermis, and a few number of melanocytes are located in mucosa. Pigment granules constituted with melanin can distribute and transport to neighboring keratinocytes [1]. Mutations in melanocortin-1-receptor (MC1R) are pivotal for human skin's tanning and pigmentation. MC1R belongs to a G-protein-coupled receptor (GPCR) that is expressed in epidermal melanocytes in a preferential manner [2]. α-Melanocyte-stimulating hormone (α-MSH), the GPCR's ligand, is a propigmentation hormone which is generated and secreted by both keratinocytes and melanocytes in the skin. After UV irradiation, α-MSH can activate GPCR. Pro-opiomelanocortin (POMC) is a multicomponent precursor for α-MSH (melanotropic), ACTH (adrenocorticotropic), and the opioid peptide β-endorphin, and α-MSH and other bioactive peptides are the cleavage products of POMC [2]. Normal synthesis of α-MSH and ACTH is extremely important to constitutive human pigmentation and the cutaneous response to UV [2].

In melanocytes, the amount and type of pigment production are regulated by MC1R. So MC1R is an important determiner of skin phototype, sensitivity to UV radiation-induced damage, and skin cancer risk [3]. The heterotrimeric G proteins consist of α, β, and γ subunits. Upon ligand binding, a signal is transmitted by GPCRs to heterotrimeric G proteins, which results in the separation of the α subunit from the Gβγ subunit of G proteins. ATP is catalyzed to be directly transformed to cAMP by the G proteins of the Gαs class and cAMP is in charge of melanogenesis including the sensitization of tyrosinase in melanin biosynthesis upon being activated by ligands such as α-MSH [4].

p53 is not only a transcriptional factor but also a tumor-suppressor protein, which is documented to directly sensitize the transcription of a lot of genes including those that control cell cycle, apoptosis, and others. POMC/MSH inducement by UV irradiation in skin is directly regulated by p53 and POMC promoter is stimulated in response to UV irradiation. p53 involves in UV-independent pathologic pigmentation and could imitate the tanning response [1]. Dipyrimidine C to T substitutions including CC to TT frameshift mutations in the p53 gene can be uniquely induced by UV in the skin of UV-irradiated mice months before tumor development [5]. In addition, p53 has been demonstrated to be necessary to the presentation of "sunburn cells," which are a sign of sunburns [5].

DUH characterized by extensively mottled pigmentation is a heterogeneous disorder, which was first diagnosed by two Japanese researchers in two generations of two pedigrees for about 80 years [6, 7]. Similar Chinese DUH pedigrees with dyschromatosis symmetrica hereditaria (DSH) with autosomal dominant DUH had been reported by us in 2003 [8] and diagnosed as DUH rather than DSH afterward. Although novel mutations of SASH1 have been identified to be associated with dyschromatosis universalis hereditaria [9], less pathogenesis of DUH has not been investigated. The pathogenesis of DUH remain unclear and indefinite for 80 years [7].

SASH1, a previously described novel family of putative adapter and scaffold proteins transmitting signals from the ligand to the receptor, was first showed to be a candidate tumor-suppressor gene in breast cancer and colon carcinoma [10–12]. Our previous study demonstrates that SASH1 interacts with Gαs, the downstream player of α-MSH/MC1R signaling pathway [13]. Our previous report indicated that in the several affected DUH individuals, hyperpigmented macules became more obvious after strong UV irradiation particularly in summer [8], but no further investigations was performed to identify the reasons of photosensitivity [14]. The significance of expression of p53/POMC/α-MSH in UV-photopigmentation response and UV-independent hyperpigmentation has been explained [1]. However, few investigations were performed to reveal that the mutations in SASH1 gene are related to hyperpigmentation and how these mutations result in hyperpigmentation phenotype.

In a word, we assume that a novel p53/POMC/α-MSH/Gαs that SASH1 involves in regulating UV-photopigmentation response and pathological hyperpigmentation phenotype.

2. Materials and methods

2.1 DUH pathologic gene sequencing and SASH1 mutation analysis

Two Chinese pedigrees recruited from the Henan and Yunnan provinces of China and one American pedigree with typical features of DUH were involved in this study. Three DUH pedigrees with an autosomal dominant inheritance pattern were diagnosed by skilled clinical dermatologists. The small American pedigree only offered peripheral blood samples of the affected individuals for investigations. This study was recognized by the ethical review committees from the appropriate institutions. Genotyping was implemented, and the two-point LOD score was counted as our previous description [8]. In total, 50 family members and 500 normal individuals (controls) involved in the research were provided with informed consent and specimens of peripheral blood DNA were acquired from all obtainable family members. PCR and sequencing were executed as our previous description [8]. ABI BigDye Terminator Cycle Sequencing Kit (Applied Biosystems, CA) was used to perform the sequencing on an ABI PRISM 3130 DNA Analyzer (Applied Biosystems) and sequence analysis software, version 3.4.1 (Applied Biosystems) were used to analyze the data. Sequencher 4.10.1 (Gene Codes Corp.) was used to compare the sequence data with SASH1 reference sequence (GenBank NM_015278.30). Nucleotide numbering reflects complementary DNA (cDNA) numbering, with +1 corresponding to the A of the ATG translation initiation codon in the reference sequence [8].

2.2 Constructing expression vectors of SASH1, Gαs, POMC, and p53

Wild-type and mutant SASH1-PEGFP-C3 and wild-type and mutant SASH1-PBABE-Flag-puro were constructed according to the protocol of our previous study [13]. To generate the p53-HA-Pcna3.0, POMC-myc-Pcdna3.0 and Gαs-Pegfp-C3 vectors, high fidelity DNA polymerase (Phusion Hot Start High Fidelity Polymerase from New England Biolabs, Inc. or GXL Polymerase from Takara) and the primers

Name of primers or probes	Sequences (5′–3′)
Gαs forward primer (SalI site included)	ACGCGTCGACATGGGCTGCCTCGGGAAC
Gαs reverse primer (XhoI site included)	CCGCTCGAGTTAGAGCAGCTCGTACTGACG
p53 forward primer (BamHI site included)	CGCGGATCCGCCACCACCATGGAGGAGCCGCAGTCAGATCCTA
p53 reverse primer (XhoI site included)	CCGCT CGAG TCAGTCTGAG TCAGGCCCTTCTGT
POMC forward primer (BamHI site included)	CGCGGATCCATGCCGAGATCGTGCTGC
POMC reverse primer (XhoI site included)	CCCAAGCTTTCACTCGCCCTTCTTGTAGGCGTTCTTGAT
SASH1 probe 1#	GCCCAAGCTTTCACACTTGTTT
SASH1 probe 2#	CCAAGACTTGCTAGAAGGAACGAGTCG
SASH1 probe 3#	CGTGGCCACCTAG ACCCGAGGTG

Table 1.
Sequences of primers or probes used in gene cloning and EMSA in this study.

indicated in **Table 1** were used to amplify the bacteria (obtained from Han Jiahuai Lab, Xiamen University, Xiamen, China) containing the vector of full-length CDS sequences of and Gαs, p53 and POMC. Mammalian expression vector (Invitrogen) via the relative restriction sites and sequenced.

2.3 Culture of cell and vector transfection

SK-MEL-28, HEK-293T, and A375 cells were cultured according to our previous description [15]. Normal human epithelial melanocytes (NHEMs, C-12402, PromoCell, Germany) were maintained in M2 medium. Lipofectamine 2000 (11668-027, Invitrogen) as previously described [15, 16] or Entranster-D (18668-01, Engreen Biosystem Co., Ltd.) or polyethyleneimine (PEI) prepared by ourselves were, respectively, used for the transfection of SK-MEL-28, A375, B16, and HEK-293T cells. The transfected A375 and SK-MEL-28 cells were screened with 1.5 μg/ml puromycin or 2.0 μg/ml G418 to construct stable cell lines. Wild-type and mutant SASH1-pEGFP-C3 or co-transfected with wild-type SASH1-Pbabe-Flag-puro and Gαs-Pegfp-C3 vectors were transiently introduced into HEK-293T cells for immunoprecipitation experiments. p53-HA-Pcdna3.0, POMC-myc-Pcdna3.0, Gαs-Pegfp-C3, and wild-type SASH1-pEGFP-C3 according to pairwise combination were introduced into NHEMs and HEK-293 or HEK-293T cells to detect the expression of exogenous p53, POMC, Gαs, and SASH1 using PEI or PromoFectin (PK-CT-2000-MAC-1, PromoCell).

Gαs-GFP, HA-p53, myc-POMC, and GFP-SASH1 recombined vector were introduced into HEK-293T cells. 24 h later, Entranster™-R transfection reagent (18668-06, Engreen Biosystem Co., Ltd) was used to transfect Gαs- and POMC-specific siRNAs synthesized by Shanghai GenePharma Co., Ltd. (Shanghai, China) to silence the expression of exogenous Gαs, p53, and SASH1 in the transfected HEK-293T cells. The sense/antisense sequences of each siRNA for Gαs, POMC are documented in **Table 2**.

Gene name	Forward primer (5′–3′)	Reverse primer (5′–3′)
SASH1	CGGGAAACGTCAAGTCGGA	ATCTCCTTTCTTGAGCTTGAG
TYRP1	CACAGGCACAGGTACCACCTC	CTGAACTACCCTAGGTCTTCGTT
Pmel17	AAGGTCCAGATGCCAGCTCAA	CTTTCACGGCTCTAGGACGTC
Rab 27a	AACTAGTGCTGCCAATGGGACA	TTTGATCGCACCACTCCTTC
Gαs	GTCCTTGCTGGGAAATCG	CGCAGGTGAAATGAGGGTAG
p53	CCACCATCCACTACAACTACAT	TCCCAGCACAGGCACAAA
POMC	AGTTCAAGAGGGAGCTGACTGG	CATGAAACCGCCGTAGCG
GAPDH	CACCCACTCCTCCACC TTTG	ACCACCCTGTTGCT GTAGCC
Gαs siRNA 1	GAGGACUACUUUCCAGAAUTT	AUUCUGGAAAGUAGUCCUCTT
Gαs siRNA 2	GCAGCUACAACAUGGUCAUTT	AUGACCAUGUUGUAGCUGCTT
POMC siRNA1	ACCUCACCACGGAAAGCAATT	UUGCUUUCCGUGGUGAGGUTT
POMC siRNA2	AGUACGUCAUGGGCCACUUTT	AAGUGGCCCAUGACGUACUTT
GAPDH	GUAUGACAACAGCCUCAAGTT	CUUGAGG CUGUUGUCAUACTT
Negative control	UUCUUCGAACGUGUCACGUTT	ACGUGACACGUUCGG AGAATT

Table 2.
Primers used for site directed mutagenesis, real time RT-PCR and RNAi.

2.4 Pull-down experiments and nanoflow LC-MS/MS and bioinformatic assays

Procedure of the pull-down assay, LC-MS/MS analyses, database search, and bioinformatic analyses for functional classification are mainly as performed as our previous description [13].

2.5 Immunoprecipitation and westernblotting

Transfected HEK-293T cells or HEK-293 cells or NHEMs with ectopic exogenous genes were washed in a gentle way for three times with PBS and lysed in IP-western blot lysis buffer (P0013, Beyond Time BioScience and Technology Company) in the presence of a complete protease inhibitor cocktail, 1 μM sodium orthovanadate, and 1 mM sodium fluoride per 10 cm dish on ice for 20 min to acquire lysisprotein. Cell lysates were centrifuged for 10 min at 12,000 rpm at 4°C. 600 μl of supernatants of cell lysates were pre-cleaned with Protein A/G PLUS-Agarose (Santa Cruz Biotechnology, Inc.) for 1 h. GFP-Tag (7G9) mouse mAb (Shanghai Abmart, Inc.) or DYKDDDDK-Flag-Tag mouse mAb (Shanghai Abmart) or HA-Tag mouse mAb (Shanghai Genomics) was used to immunoprecipitate the corresponding proteins at 4°C for 2 h. Then, the cell lysates were mixed with 20 μl of Protein A/G PLUS-Agarose (Santa Cruz Biotechnology, Inc.) at 4°C for 10 h for co-immunoprecipitation or immunoprecipitation analyses. Finally, immunoprecipitates were washed for three times with PBS and subjected to western blotting. GFP-Tag mouse Ab, Flag-tag mouse mAb, DYKDDDDK-Flag mouse mAb, GAPDH mouse mAb and anti-β-tubulin mouse mAb (Shanghai Abmart, Inc.), anti-Gαs rabbit polyclonal Ab (Gene Tex, Inc.), myc-tag mAb and HA-tag mouse mAb (Shanghai Genomics), SASH1 Rabbit mAb (Bethyl Laboratories, Inc.), TYRP1 (TA99) mouse mAb and melanoma gp100 Rabbit mAb (Abcam), Rab 27a mouse mAb (Abnova) were used for immunoblotting assay as previously described [17, 18].

2.6 Immunohistochemical analyses and immunofluorescence staining and melanin staining

2.6.1 Immunohistochemistry

All participating patients in this study were given the written informed consent regarding tissue and data use for scientific purposes. Epithelial tissues from affected individuals with the Y551D SASH1 mutation from pedigree I were fixed and embedded in paraffin. Paraffin sections (5 μm) were baked at 56°C for 3 h, and then deparaffinized and rehydrated using xylene and an ethanol gradient. SASH1 monoclonal antibody, rabbit anti-ACTH antibody, MiTF polyclonal antibody, the antibodies of melanogenesis related molecules including HMB45, TYRP1, and Rab 27a and p53 monoclonal antibody was used to bind the corresponding proteins on paraffin sections, respectively. Finally, counterstaining of hematoxylin was performed and the sections were photographed under the positive position microscope BX51.

2.6.2 Immunofluorescence (IF) and confocal microscopy detection

A375 stable cells with ectopic wild-type or mutant SASH1 in 6-well chamber slides were analyzed with indirect immunofluorescence analysis. SASH1 rabbit mAb (Betheyl Laboratories, Inc.) and DYKDDDDK-Flag mouse mAb (Shanghai Genomics, China) were used to assess SASH1 localization and expression, as described previously [13].

2.6.3 Melanin staining

The melanin staining of paraffin sections obtained from epithelial tissues were performed as our previous descriptions and observed under a light microscope [18].

2.7 Quantitative RT-PCR

TRIzol reagent (Invitrogen) was used to isolate the total RNA from the different groups of SK-MEL-28 cells. Reverse transcription was performed according to the manufacturer's protocol for the PrimeScript™ RT Reagent Kit (DRR037A, TaKaRa) for qRT-PCR. The sense and antisense primer sequences for SASH1, TYRP1, Pmel17, Rab27a, Gαs, POMC, and GAPDH are showed in **Table 2**. The PCR products were identified by agarose gel electrophoresis. The Applied Biosystems 7500 System was applied to detect the expression of corresponding genes with SYBR Premix Ex Taq™ (DRR041A, TaKaRa). The quantity of each mRNA was normalized to that of GAPDH mRNA.

2.8 UV irradiation

Human foreskin tissues from a 14 year-old boy were irradiated for enough time under a ultraviolet phototherapy instrument (NBUVB SS-05, Sigma) to reach the expected UV intensity. The irradiated tissues were fixed in 10% formalin and embedded in paraffin for immunohistochemistry analyses. We conformed to the guidelines of the World Medical Assembly (Declaration of Helsinki) to acquire the human foreskin tissues. In vitro UV experiments were mainly referred to the protocol of our institute [19]. HEK-293T cells and NHEMs transiently with ectopic myc-POMC were subcultured to approximately 70–80% confluence and were irradiated with 100 mJ/cm^2 UVC delivered via a HL-2000 HybriLinker with a 254 nm wavelength and followed by the indicated recovery time. Finally, immunobloting was performed to identify the corresponding proteins' levels.

2.9 Electrophoretic mobility shift assay

Three probes binding with/without biotin targeting the promoter sequence of SASH1 gene were synthesized. The sequences of probes were as indicated in **Table 1**. Electrophoretic mobility shift assay was performed as described as the protocol provided with LightShift® Chemiluminescent EMSA (20148, Thermo Scientific, Pierce Biotechnology) to detect the bindings of SASH1 with p53 [18].

2.10 Statistical analyses

The data are indicated as mean ± standard error of the mean (SEM)s. The homogeneity of variance test was first used to analyze the variance homogeneity of data and the data were analyzed the change of variable test. Statistical significance was determined by a one-factor analysis of variance (ANOVA) using LSD on SPSS version 16.0 to produce the required p-values. Cartograms were plotted with GraphPad Prism 5.

3. Results

3.1 Mutations in SASH1 gene in the DUH-affected individuals result in up-regulated SASH1 in vitro and in vivo

The gene that is responsible for DUH had been localized to chromosome 6q24.2-q25.2. The 10.2 Mb region on chromosome 6 (6q24.2-q25.2) containing more than 50 candidate genes is flanked by the markers D6S1703 and D6S1708 [8]. Direct sequencing of the PCR products of exons amplified from genomic DNA of affected, unaffected, and control individuals was performed to screen the selected genes in this region for possible pathological mutations. 50 candidate genes were sequenced. Finally, in the probands in each of the two nonconsanguineous Chinese DUH-affected pedigrees (families I and II) and in one nonconsanguineous American DUH-affected pedigree (family III), three heterozygous mutations encoding amino acid substitutions in SAM and SH3 domain containing I (SASH1) were found in the three pedigrees. The substitution mutations in SASH1 gene were as follows: a TAC → GAC substitution at nucleotide 2126 in exon 14, causing a Y551D (p.Tyr 551 Asp) mutation at codon 551 in family I, a CTC → CCC substitution at nucleotide 2019 in exon 13, causing a L515P (p.Leu to Pro) mutation at codon 515 in family II, and a GAA → AAA substitution at nucleotide 2000 in exon 13, resulting in a E509K (p.Glu to Lys) mutation in family III. These sequence alterations were identified in all of the affected pedigree members, but were not observed in unaffected pedigree members, correlating the presence of the mutations with the presence of the phenotype. In any of the 500 normal controls or in any of the current databases, including the HapMap database, these three SASH1 mutations were not found [18]. So, these three mutations are impossible to be common single nucleotide polymorphisms (SNPs) [8].

In A375 stable cells with ectopic SASH1 gene mutants, mutated SASH1 were found to be significantly up-regulated (**Figure 1B**). Western blot showed that up-regulation of SASH1 was found in A375 cells stably expressing either wild-type (WT-A375 cells) or mutant SASH1, when compared to the expression of endogenous SASH1 in A375 cells with expression of pBABE-puro empty vector (VECTOR-A375 cells) or BLANK-A375 cells (**Figure 1B**) [18].

To identify the stability of SASH1 proteins, 20 μg/ml of the protein synthesis inhibitor cycloheximide (CHX) was used to treat HEK-293T stable cells with ectopic wild-type or mutant SASH1 expressing for the indicated times to assess the half-life of SASH1. SASH1 protein levels were induced to decrease by CHX treatment in a time-course-dependent manner. Wild-type SASH1 levels was decreased with a half-life of approximately 4 h, however, mutant SASH1 proteins began to degrade with CHX treatment for 6 h or longer. Therefore, it is deduced that the three mutant SASH1 proteins were more steady than the wild-type, confirming the conclusion offered above that expression levels of mutated SASH1(s) are higher levels than that of wild-type (**Figure 2A** and **B**). Endogenous SASH1 was not a stable protein with an half-life of approximately 3 h as identified by western blot (**Figure 2C**) [18].

The subcellular localization of SASH1 in A375 cells and skin epithelial layers was further characterized. A homogenous pattern of expression of SASH1 protein was observed in VECTOR-A375 cells and the skin epithelial layers from normal controls (**Figures 1C** and **3-a4**). However, heterogeneity expression of SASH1 protein was showed in WT-A375 cells and mutant-A375 cells (**Figure 3-b4–e4**) as well as in the epithelial tissues of affected individuals (**Figure 1C**). In addition, as identified by Mitf, a melanocyte indicator, most of the SASH1-positive cells were Mitf-nucleic positive-melanocytes in the epithelial tissues of DUH-affected individuals. These Mitf-nucleic positive-melanocytes in the affected epithelial layer showed a heterogeneous

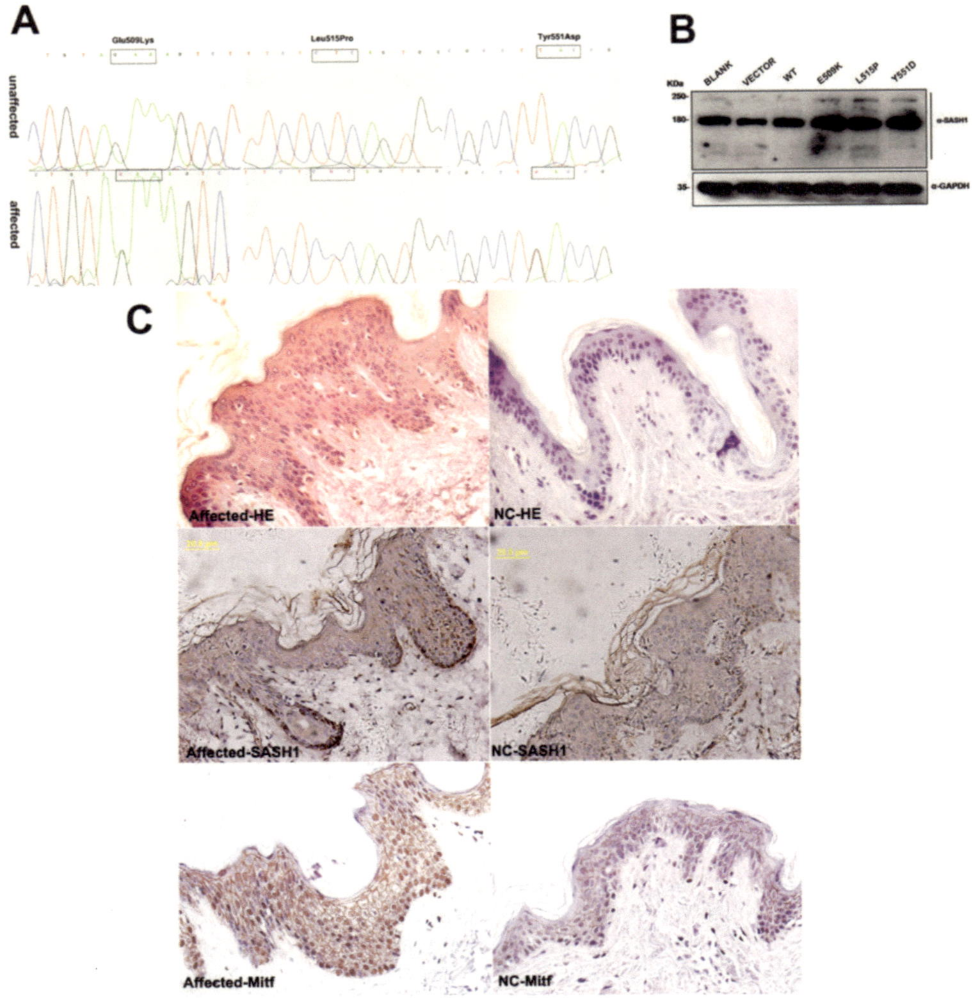

Figure 1.
Increased SASH1 expression is induced by mutations in SASH1 in vitro and in vivo. (A) Substitution mutation sites in the SASH1 gene in three DUH pedigrees. (B) Differential and increased expression of mutant SASH1 proteins is detected compared to that of wild-type SASH1 in different A375 cells by immunoblot. (C) Immunohistochemistry detection of SASH1 and Mitf. Heterogeneous SASH1 protein was detected in all of the DUH-affected epithelial layers compared to that of normal controls (NC). Heterogeneous distribution of melanocytes is detected in the epithelial layers of DUH-affected individuals using Mitf compared to that of normal controls. 400× magnification. Scale bar = 20 μm. The representative positive cells of SASH1 and Mitf were denoted by red arrows.

distribution compared to those of unaffected individuals (**Figure 1C**). Some Mitf-tenuigenin-positive staining is of false positivity (**Figure 1C**). Melanocytes or SASH1-positive epithelial cells not only localized at the basal layers, but also the suprabasal layers of the affected epidermal tissue, the phenomenon of which coincides with our previous descriptions that SASH1 mutations promotes melanocyte movement from the affected basal layers to the superficial ones [13].

3.2 SASH1 binds to Gαs and is induced by the canonical p53/POMC/Gαs cascade

SASH1 contains two functional domains, SAM and SH3 domain, which indicates SASH1 may plays an important role in a signaling pathway acting as a signaling molecule adapter or an associated scaffolding protein [8, 9]. Therefore, we performed a pull-down assay and a mass spectrometry analysis to investigate which signaling

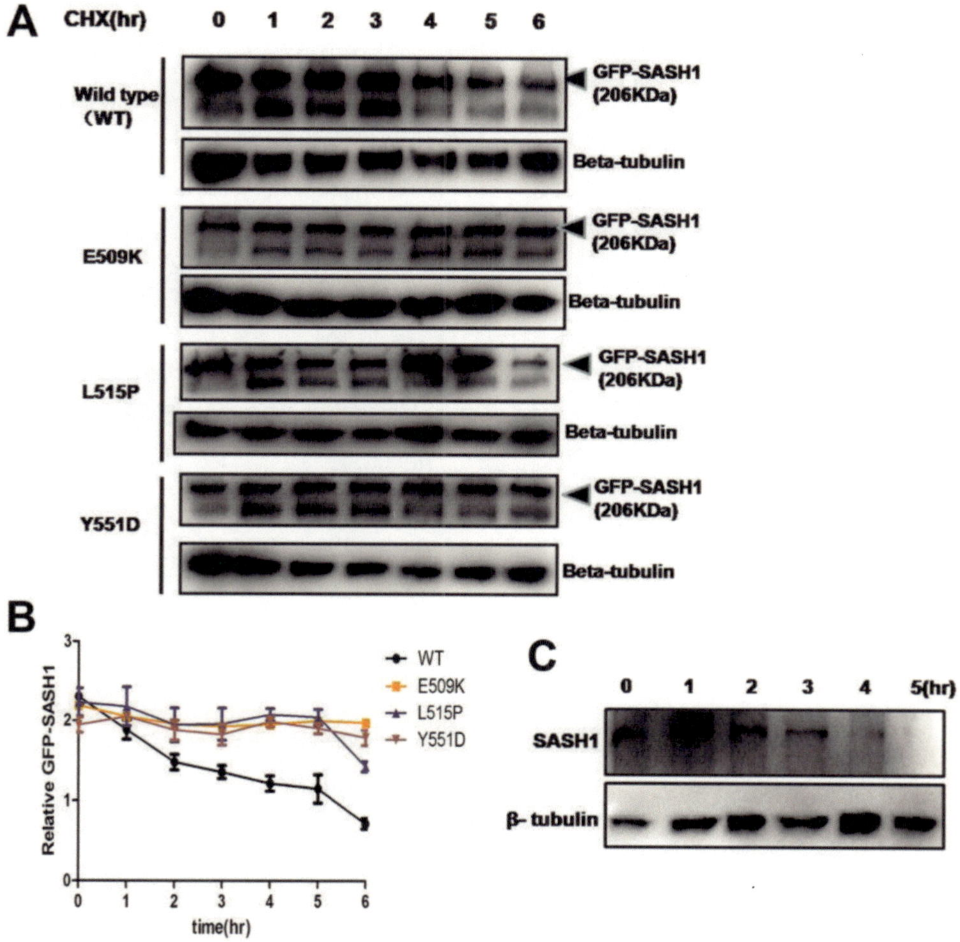

Figure 2.
Endogenous SASH1 protein is unstable and mutation of SASH1 induces the heterogeneous expression of SASH1 in vitro. (A) Mutant SASH1 proteins are more stable than the wild-type SASH1 protein. Stable HEK-293T cells were treated with CHX (20 μg/ml) for the indicated times and analyzed by western blotting. The amount of SASH was quantified by densitometry and normalized to β-tubulin. CHX resulted in the degradation of wild-type SASH1 protein, which had a half-life of 4 h. Under a 6-h or longer treatment with CHX, CHX began to induce the degradation of mutant SASH1 proteins. (B) The intensity of GFP-SASH1 was quantified by densitometry and normalized to β-tubulin (n = 3). (C) Endogenous SASH1 is an unstable protein. HEK-293T cells were deprived of FBS for the indicated time and lysed and subjected to western blot to detect the endogenous SASH1 levels.

pathways are regulated by SASH1. Pull-down experiments and nanoflow LC-MS/MS analysis demonstrated that SASH1 interacts with Gαs and CALM, both of which are important in melanogenesis process (**Tables 3** and **4**) in WT-A375 cells. Gαs connects receptor-ligand associations with the activation of adenylyl cyclase and many cellular responses, serving as a pivotal player in the conventional signal cascades [20]. To investigate the associations between SASH1 and Gαs, HEK-293T cells were co-transfected with Flag-SASH1 and GFP-Gαs. Exogenous SASH1 was immunoprecipitated with both exogenous Gαs (GFP-Gαs) and endogenous Gαs. Immunoprecipitates of exogenous SASH1 had different observed band sizes of Gαs (**Figure 4B** and **C**) [18].

Gαs mediates cAMP production in melanocytes which is stimulated by α-MSH and melanocortins [21] and our study here shows that Gαs is associated with SASH1. Hence, we examine whether Gαs is required for the induction of SASH1 and how Gαs mediates SASH1 expression, we introduced exogenous p53, POMC, Gαs, and SASH1 gene into HEK-293T and NHEMs (normal human epithelial

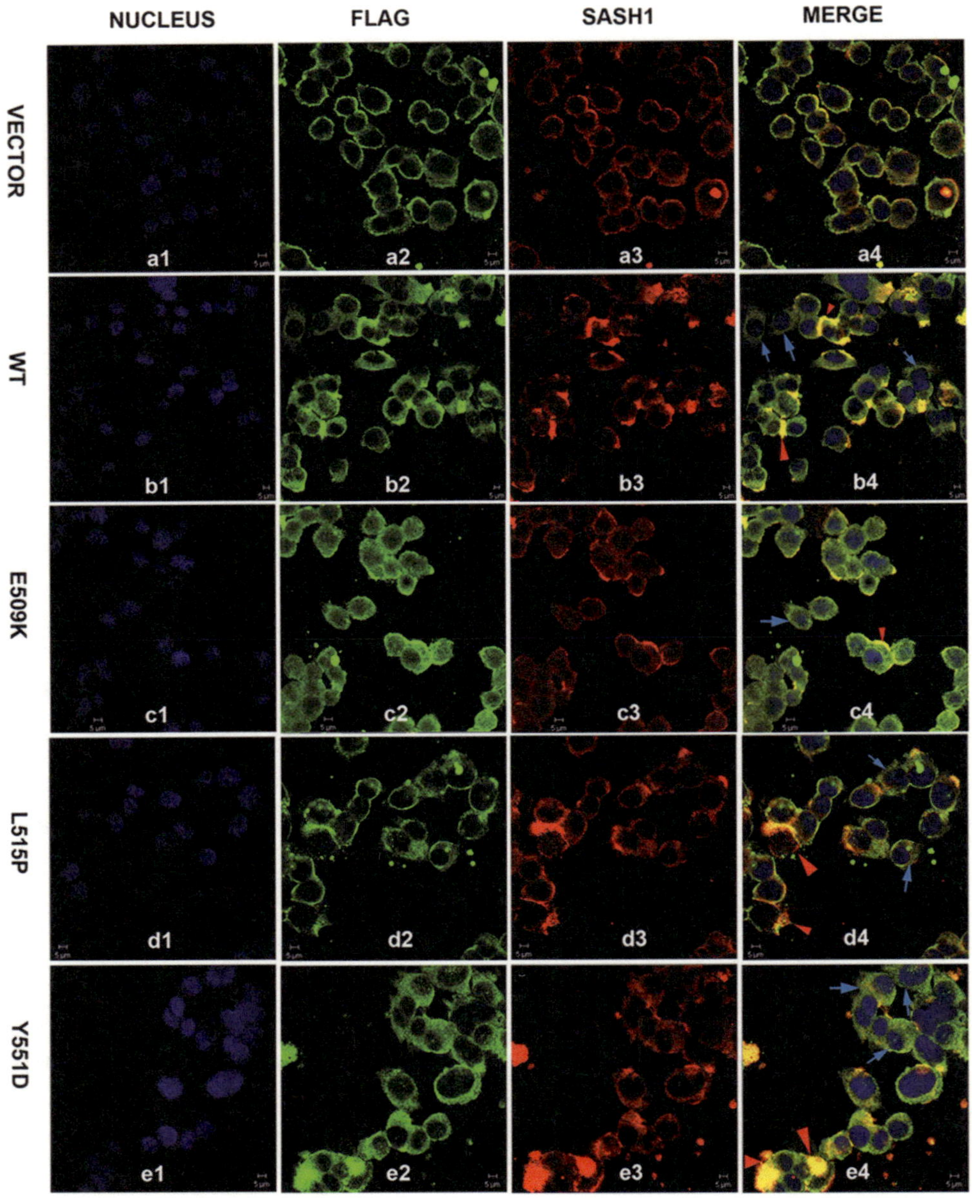

Figure 3.
*Subcellular localization of SASH1. The fluorescence signals that were detected by confocal microscopy indicate that the overexpression or mutation of SASH1 results in the heterogeneous expression of SASH1 in vitro in A375 stable cells. The green fluorescence represents the Flag-tag label. Both exogenous and endogenous SASH1 are labeled with a red fluorescent tag. The nuclei are labeled with DAPI (in blue). The yellow fluorescence indicates the overlap of the green and red fluorescent staining. The red arrowheads indicate the activated SASH1-Flag fusion proteins that were expressed in the cytoplasm of WT-A375 cells or mutant-A375 cells. The blue arrowheads indicate the regions that do not express the activated SASH1-Flag fusion protein in the cytoplasm of WT-A375 or mutant-A375 cells. The endogenous SASH1 presents a uniform pattern of expression in all of the VECTOR-A375 cells (*__Figure 3A-a4__*). Scale bar = 5 μm.*

melanocytes) to assess the effects of p53 and POMC on Gαs. Exogenous Gαs was induced in the co-existence of exogenous p53 and POMC (**Figure 4C** lane 5 and **Figure 4D** lane 5) and both inducements of exogenous Gαs and exogenous SASH1 were observed in the co-existence of exogenous p53 and POMC in two types of normal cells (**Figure 4C** lane 6 and **Figure 4D** lane 6). Meanwhile, in the presence of GFP-SASH1, GFP-Gαs was also induced (**Figure 4C** lane 4 and **Figure 4D**

Protein name	Score	Protein possibility	Total peptide	Unique peptide
SASH1	200.3	1	37	20
Gαs	20.2	1	8	5
CALM1	10.2	1	7	3

Affinity-purified proteins were identified by MS analysis and the detailed peptide sequences are summarized in **Table 2**.

Table 3.
Proteins interacting with SASH1 were identified by MS analysis.

Protein name	Peptide sequence
SASH1 (O94885)	K.KPSTEGGEEHVFENSPVLDERS R.AVLLTAVELLQEYDSNSDQSGSQEKL K.GEDVGYVASEITMSDEERI R.VSQDLEVEKPDASPTSLQLR.S R.VHTDFTPSPYDTDSLKI K.LLEEEDLDELNIRD K.LHAEGIDLTEEPYSDKH K.PGAGTSEAFSR.L KPLFFDGSPEKPPEDDSDSLTTSPSSSSLDTWGAG K.MGTFFSYPEEEKA KMITIEEALARL RSLHVGSNNSDPMGKE SLHVGSNNSDPMGK ITIEEALAR MITIEEALARL RGVDLETLTENKL IPSQPPPVPAK TIEEALAR KYFWQNFRK SALYSGVHK
Gαs (P63092)	EAIETIVAAMSNLVPPVELANPENQFR YTTPEDATPEPGEDPR IEDYFPEFAR MFDVGGQR VLTSGIFETK
CALM1 (P62158)	R.EADIDGDGQVNYEEFVQMMTAK

A375 stable cells with ectopic SBP-FLAG-SASH1 expressing were lysed, immunoprecipitated with SBP beads and digested with trypsin. The liquid supernatant was collected, dried, and dissolved in 10% (v/v) acetonitrile and 0.8% formic acid solution. Nanoflow LC-MS/MS analyses were performed to identify the peptides.

Table 4.
SBP-FLAG-SASH1 affinity purification was performed to identify the peptide sequences of the binding complex of SASH1 protein.

lane 4), which indicated that SASH1 is necessary for activation of GFP-Gαs. And immunoblot showed that Gαs was identified to be induced by exogenous p53 and SASH1 (**Figure 4E** and **F**). Our results also demonstrated that POMC was mediated by p53 in HEK-293T and melanocytes, which were consistent with previous conclusions [1] (**Figure 4G** and **H**).

To confirm the fact that POMC, p53, and Gαs are needed for the induction of SASH1 and exogenous POMC, p53, Gαs, and SASH1 were transfected into HEK-293T cells and followed by silence of Gαs and POMC by two specific pairs of siRNA, respectively. As identified in HEK-293 cells, deletion of Gαs gene directly induced significant reduction of SASH1 (**Figure 5C** and **D**). Deletion of POMC by siRNA resulted in the downregulation of Gαs and SASH1 (**Figure 5E** and **F**). Taken above,

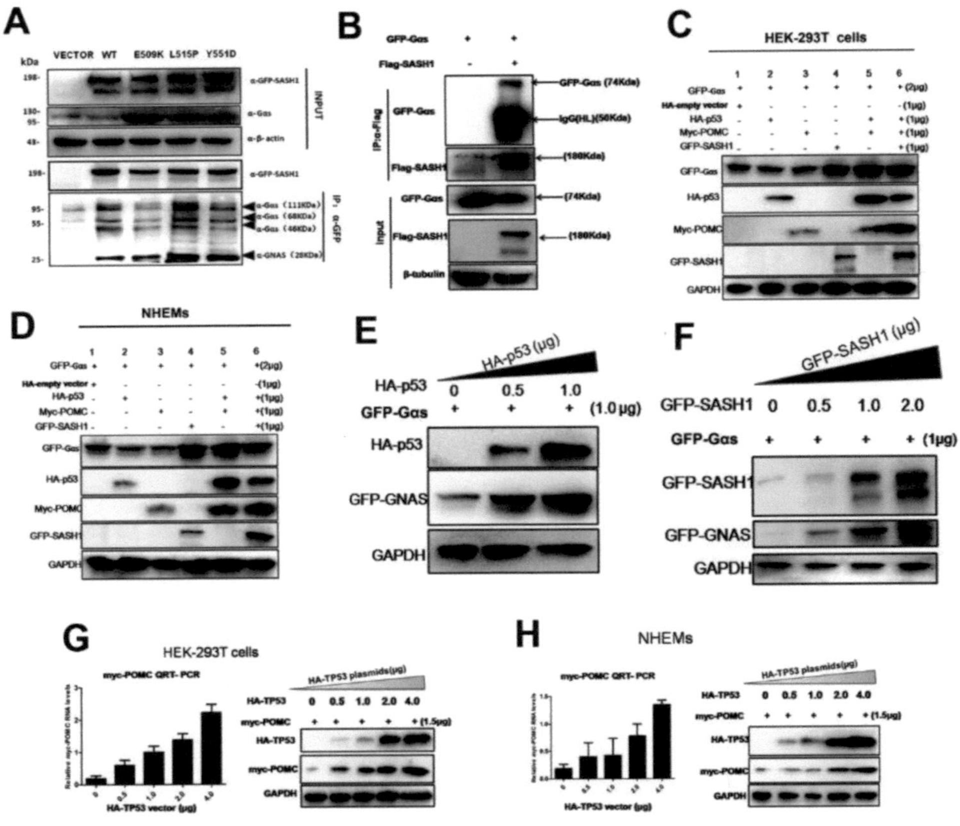

Figure 4.
Gαs binds to SASH1 and is a central downstream player of p53/POMC cascade. (A) Immunoprecipitation-Western blot (IP-WB) was performed to identify the interactions between GFP-SASH1 and endogenous Gαs in HEK-293T cells. pEGFP-C3-SASH1-recombined vectors were introduced into HEK-293T cells. At 24 h after transfection, GFP-SASH1 was immunoprecipitated and the associated endogenous Gαs was identified by immunoblot analysis using a Gαs antibody. (B) Exogenous Gαs binds to exogenous SASH1. pEGFP-C3-Gαs and pBABE-puro-Flag-SASH1 vectors were co-introduced into HEK-293T cells. At 36 h after transfection, exogenous SASH1 (Flag-SASH1) was immunoprecipitated and the associated GFP-Gas was detected by western blot analysis using an anti-GFP antibody. At 36-h post-transfection, Flag-SASH1 was immunoprecipitated and the associated exogenous Gαs (GFP-Gas) was detected by immuoblot using an anti-GFP antibody. (C) and (D) P53, POMC, and SASH1 are essential to the activation of Gαs. HA-p53, myc-POMC, and GFP-SASH1, respectively, according to different manners of combination were introduced into HEK-293 cells and NHEMs. After 36 h after transfection, immunoblotting was performed to detect the protein levels in two normal cells along with GAPDH as loading control. (E) Exogenous Gαs (GFP-Gαs) is induced by exogenous p53 (HA-p53). HEK-293 cells were transfected with HA-p53 and GFP-Gαs. At 36-h post-transfection, the transfected HEK-293 cells were lysed and subjected to western blot analyses. GFP-Gαs was induced by gradually increased amounts of HA-p53. (F) Exogenous Gαs (GFP-Gαs) is induced by exogenous SASH1 (GFP-Gαs). GFP-Gαs and GFP-SASH1 were transfected into HEK-293T cells. GFP-Gαs was induced by gradually increased doses of GFP-SASH1. (G) and (H) Exogenous POMC (myc-POMC) was induced by increased dose of exogenous p53 in HEK-293T cells and NHEMs. Different amounts of HA-p53 vector and a certain amounts of myc-POMC vector were introduced into HEK-293T cell for expression. Exogenous POMC RNA levels were quantified by quantitative RT-PCR and normalized to GAPDH. The expression of HA-p53 and myc-POMC was analyzed by immunoblot using GAPDH as loading control.

it is believed that Gαs serves as a pivotal downstream of p53/POMC cascade and SASH1 is regulated by a novel p53/POMC/Gαs cascade.

3.3 P53 physiologically triggers SASH1 upon UV irradiation

To reveal the phenomenon that P53 physiologically triggers SASH1, discarded normal human foreskin samples were irradiated to gradually increased dose of UV and stained for the histological analyses of p53, ACTH/POMC, and SASH1.

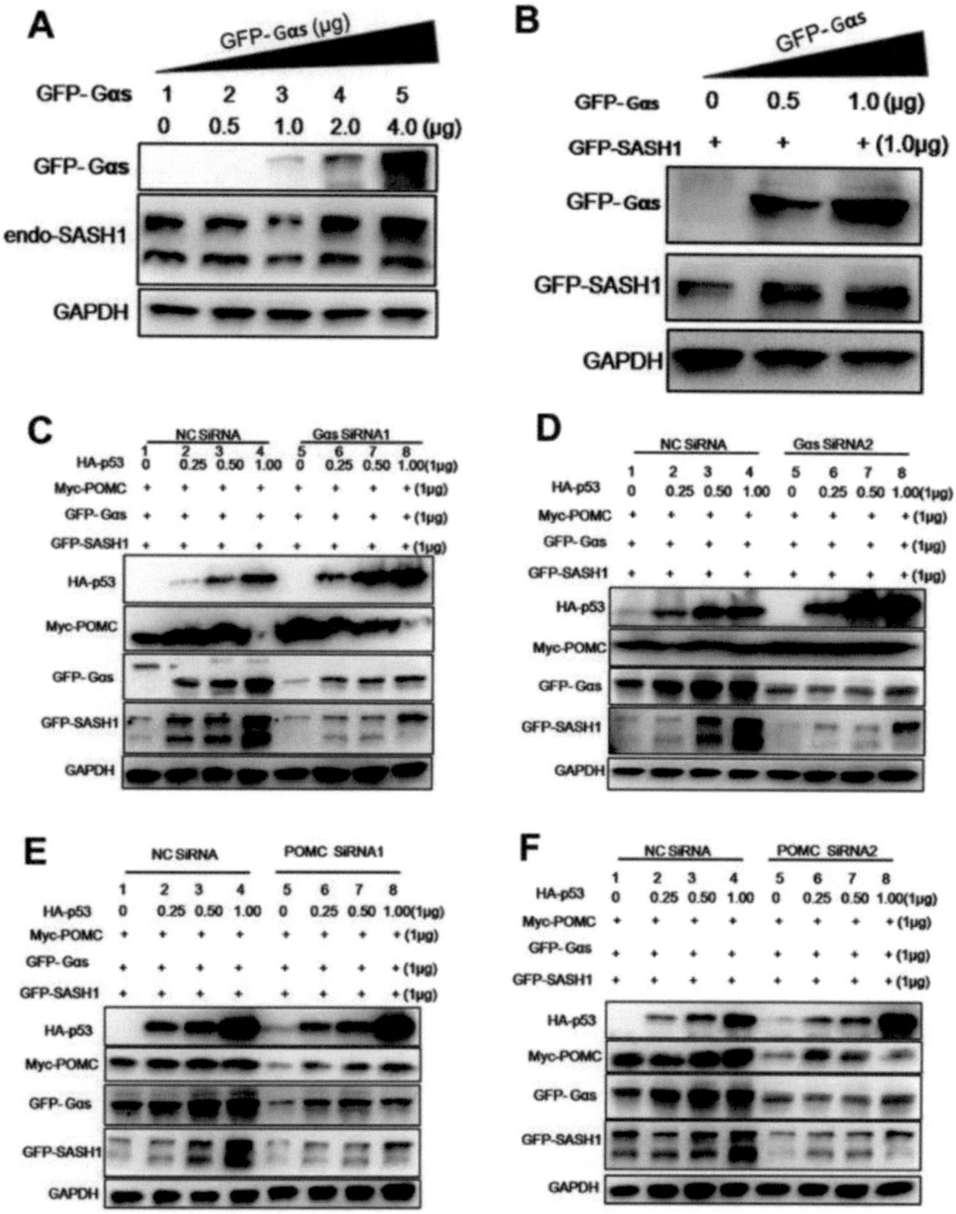

Figure 5.
Expression of SASH1 was mediated by a novel p53/POMC/Gαs/SASH1 signal cascade. (A) and (B) Endogenous and exogenous SASH1 (GFP-Gαs) are both induced by Gαs. HEK-293T cells were transfected with gradually increasing doses of exogenous Gαs and exogenous SASH1, or only different amounts of exogenous Gαs. The protein levels of endogenous or exogenous SASH1 were assessed by immunoblot. (C) and (D) Gαs is essential for the induction of SASH1. Exogenous Gαs, POMC, and SASH1 as well as increasing doses of HA-p53 according to different combinations were transfected into HEK-293 cells. Among the transfected HEK-293 cells, two groups of cells were afterward transfected with two pairs of effective Gαs siRNAs and negative control (NC) siRNA. The corresponding protein levels were assessed by western blot. (E) and (F) POMC is essential for the induction of SASH1 and Gαs. HEK-293 cells were transfected with GFP-Gαs, myc-POMC, and GFP-SASH1 as well as increasing dose of HA-p53 according to different manner of combinations. Among the transfected HEK-293 cells, two groups of HEK-293 cells were later silenced with two pairs of effective POMC siRNAs and NC siRNA.

Immunohistochemical (IHC) analyses demonstrated p53 is quickly induced in basal layers at the 0.5 J/cm^2 dose of UV exposure. The quick inducement of SASH1 and POMC/ACTH at UV irradiation 1.0 J/cm^2 dose in melanocytes is followed closely by p53 up-regulation (**Figure 6A**). Previous study had indicated that up-regulated POMC gene is induced at both protein and mRNA levels following UV exposure of skin [22, 23]. Followed the previous reports [1], a 100 J/m^2 UVB dose was administered in this study. The 100 J/m^2 UVB dose equates to the standard erythema dose (SED), which is commonly used as a measure of sunlight [24]. Therefore, both endogenous p53 and SASH1 protein levels in HEK-293T cells and NHEMs with ectopic exogenous POMC after UV irradiation were assessed by immunoblot. Expression of exogenous POMC and endogenous SASH1 was markedly induced by 6 h after UV irradiation, which accords with its known stabilization by UV in NHEMs. At 24 h, in NHEMs, UV irradiation maximally promoted the expression of POMC, p53, and SASH1 protein (**Figure 6B**). Similar induction of exogenous POMC and endogenous p53 and SASH1 was detected in HEK-293T cells after UV irradiation (**Figure 6C**). Hence, it is believed that both POMC and SASH1 serve as novel downstream players which respond to p53 inducement by UV irradiation.

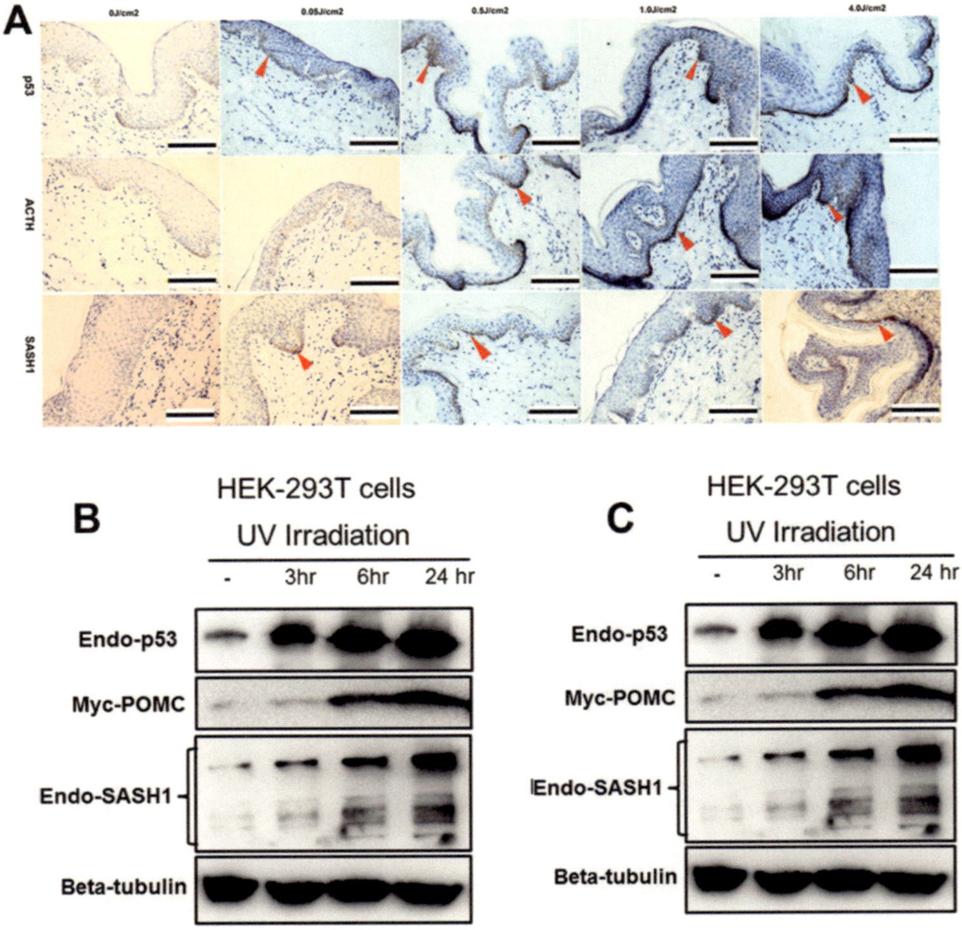

Figure 6.

SASH1 is induced physiologically by p53 after UV irradiation. (A) Upon UV irradiation or without UV irradiation, immunohistochemistry analyses of p53, POMC, and SASH1 in human foreskin indicated that p53 is activated by UV-induced-increase of POMC and SASH1. The human foreskin tissues obtained from a 14-year-old boy were irradiated at different doses of UV intensity, then fixed in 10% formalin and embedded in paraffin for immunohistochemical analyses. Scale bar: 20 μm. The representative positive cells of p53, ACTH, and SASH1 were donated by red arrows. (B) and (C) NHEMs and HEK-293T cells with ectopic exogenous POMC (myc-POMC) expression were irradiated with UV irradiation (100 mJ/cm^2) and recovered for the indicated times. Transfected cells were lysed and at different time-points after irradiation as indicated. Western blot was used to detect the protein levels of endogenous p53, endogenous SASH1, and exogenous POMC along with GAPDH or beta-tubulin as a loading control.

3.4 p53-SASH1 reciprocal inducement in normal cells

To assess whether p53 is required for the inducement of SASH1, exogenous p53, and POMC gene were transfected into HEK-293T and NHEMs to test the induction of p53 and POMC to SASH1. In NHEMs and HEK-293T cells with ectopic of POMC (myc-POMC) in NHEMs and HEK-293T cells, exogenous SASH1 were induced to up-regulate by p53 (**Figure 7**). Increased protein levels of exogenous SASH1 was induced by increasing amounts of exogenous p53 in two normal cells (**Figure 8A** and **B**). On the contrary, exogenous p53 (HA-p53) was also triggered by increasing amounts of exogenous SASH1 (**Figure 8C** and **D**). The induction of exogenous SASH1 to endogenous p53 was also identified. It has been documented that, in two types of normal cells, increased endogenous p53 was induced by increasing doses of exogenous SASH1 (**Figure 8E** and **F**).

Since SASH1 is mediated by p53, we want to investigate whether there is a direct relationship between SASH1 and p53. As indicated in **Figure 9A** and **B**, HA-p53 did not bind to GFP-SASH. So, we tested the proximal 1 kb promoter region of the SASH1 gene to find the consensus transcription-factor-binding elements that are conserved between human, rat, and mouse. Among the various consensus elements searched for, p53 gene was remarkable. A most possible p53-binding site, sequence of which is "tgcccaagctttcacacttgttt" was identified in the SASH1 5′ flanking region about 550 bp upstream of the transcription initiation site in humans (**Figure 9C**). So, three synthesized probes were used to investigate the associations of p53 protein with SASH1 gene promoter. However, analyses of electrophoretic mobility shift assay (EMSA) revealed that there was no p53 protein bind the promoter region of the SASH1 gene (**Figure 9D**).

In summary, it is believed that SASH is regulated by the p53/POMC/α-MSH/Gαs signal cascade and p53/POMC/α-MSH/Gαs cascade and SASH1 constitute a novel autoregulatory loop. The p53/POMC/α-MSH/Gαs/SASH1 regulatory loop acts as an auto-feedback circuit to regulate the p53-SASH1 reciprocal inducement (**Figure 8G**).

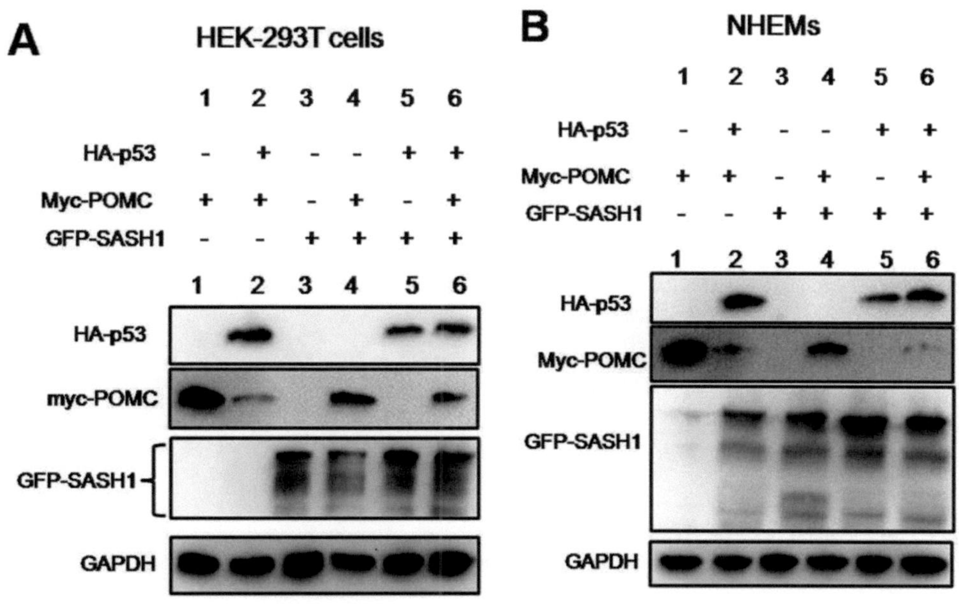

Figure 7.
Exogenous p53 triggers expression of SASH1. (A) Exogenous p53 caused up-regulation of exogenous SASH1 in HEK-293T cells. HA-TP53, GFP-SASH1, and myc-POMC were transfected into HEK-293T cells for transient expression. Cells were lysed in 0.5% NP40 buffer containing protease inhibitors and subjected to western blot along with GAPDH as loading control. (B) Exogenous p53 caused up-regulation of exogenous SASH1 in NHEMs.

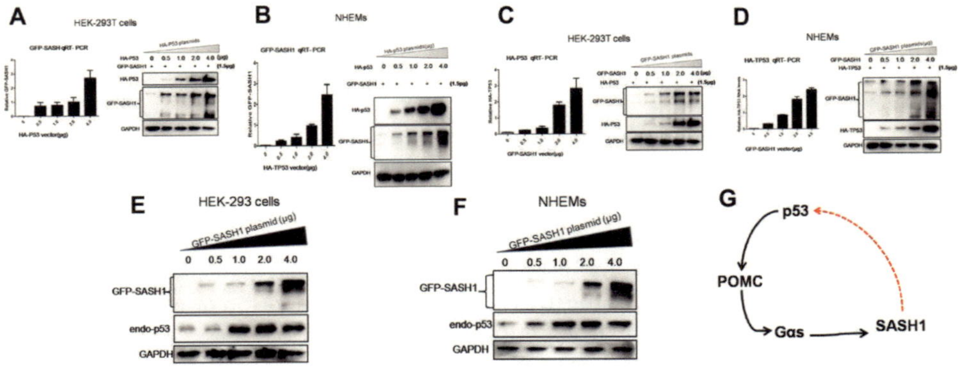

Figure 8.

There is a reciprocal induction between p53 and SASH1 in normal cells. (A) and (B) Exogenous SASH1 was triggered by exogenous p53 (HA-p53) in a dose-dependent manner. Different amounts of HA-TP53 plasmid were introduced into HEK-293T cells and NHEMs as indicated. After 48-h post-transfection or transfection, total RNA of HEK-293T cells and NHEMs was extracted and exogenous SASH1 RNA levels were assessed by quantitative RT-PCR and normalized to GAPDH. Expression of exogenous p53 protein and SASH1 were analyzed by western blot along with GAPDH as a loading control. (C) and (D) Protein and RNA levels of exogenous p53 were promoted by exogenous SASH1 promotes expression. Different amounts of GFP-SASH1 plasmid and a certain amount of exogenous p53 were transfected to HEK-293T cells and NHEMs cells. As revealed by QRT-PCR and western blot, enhanced expression of exogenous TP53 was induced by increasing amounts of GFP-SASH1. (E) and (F) Increased endogenous p53 was induced by exogenous SASH1. Different amounts of GFP-SASH1 were transfected in to HEK-293T cells and NHEMs. At 36 h after transfection, cells were lysed and subjected to western blot to analyze the expression of GFP-SASH1 with GAPDH as loading control. Results are the representative of three independent experiments. (G) A novel reciprocal induction of p53 and SASH1 is mediated by an autoregulatory p53/POMC/Gαs/SASH1 loop. p53 is activated by different types of stress, which fosters POMC, Gαs, and SASH1 successively. The inducement of SASH1 by p53/POMC/Gαs cascade promotes the up-regulation p53 in nucleus, then induced nucleic p53 conversely activates POMC/Gαs/SASH1 cascade, which consists an autoregulatory p53/POMC/Gαs/SASH1 loop.

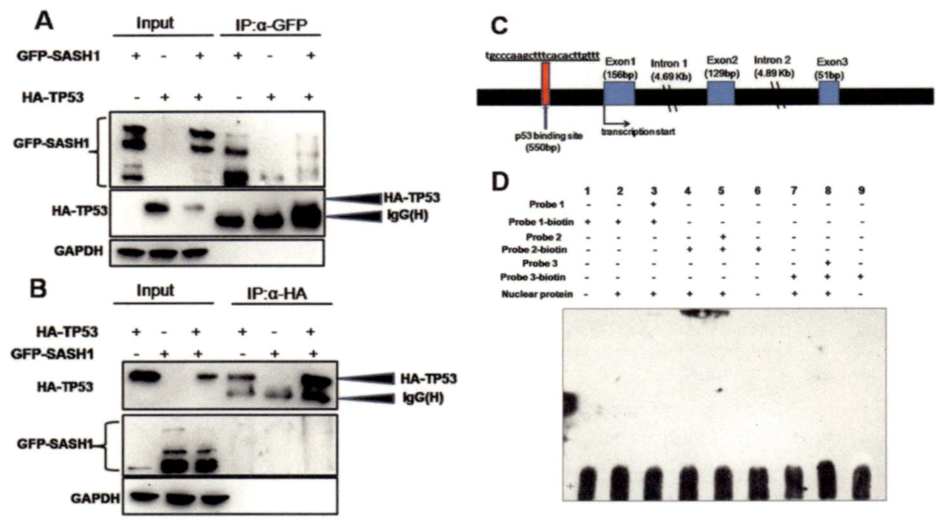

Figure 9.

p53 is not associated with SASH1 and SASH1 is not transcriptionally regulated by p53. (A) and (B) HEK-293T cells were co-transfected with the pEGFP-C3-SASH1 and Pcdna 3.0-HA-p53 vectors. At 24-h post-transfection, GFP-SASH1 was immunoprecipitated and the associated HA-p53 was detected by western blot analysis using an anti-HA antibody. Similarly, HA-p53 was immunoprecipitated and the associated GFP-SASH was detected by western blot analysis using an anti-GFP antibody. (C) Showed a schematic representation of the SASH1 locus, which indicates location of a p53-binding consensus sequence. (D) EMSA analyses demonstrated that there was none of among three probes of SASH1 gene promoter to bind p53 recombined protein.

3.5 Enhance expression of p53 and POMC is induced by SASH1 mutations

SASH1 up-regulation is mediated by SASH1 mutations, which is unfathomable enigma to us for lone time. Therefore, HEK-293T cells and NHEMs were transfected with wild-type or mutant SASH1 (wt SASH1 or mut SASH1), exogenous p53 and exogenous POMC to assess the effects of SASH1 mutations on p53 and POMC. As demonstrated in **Figure 10A** and **B**, increased expression of p53 and POMC was induced by SASH1 mutations. The effects of SASH1 mutations on endogenous p53 at protein level were also assessed. Increased endogenous p53 was also induced by mutated SASH1 (**Figure 10C** and **D**). In order to identify that p53 is induced by SASH1 mutations in vivo, immunostaining of p53 in the affected epithelial tissues with SASH1 Y551D mutation was performed. IHC analyses indicated that more nucleic expression of p53 in epithelial tissues and more p53-positive cells in affected epithelia layers were induced by SASH1 Y551D mutation (**Figure 10E**).

All of these indicate that not only SASH1 is positively regulated by the p53/POMC/α-MSH/Gαs/SASH1 autoregulatory loop, but also SASH1 mutations serve more as molecular rheostats rather than an on-off switch to control this regulatory loop.

3.6 Expression of melanosomes matrix molecules was triggered by SASH1 mutations

Since there is a SASH1-p53 autoregulatory loop, the changes of downstream partners of SASH1 need to be tested. Therefore, we further identified the effects

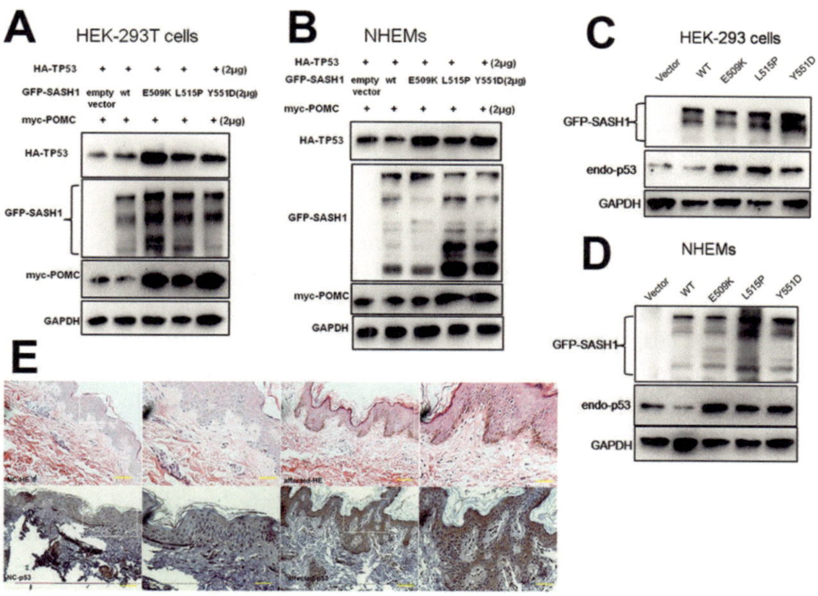

Figure 10.
p53 and POMC are induced to be increased SASH1 mutations. (A) and (B) Up-regulated SASH1 induced by SASH1 mutations promotes the expression of exogenous p53 and exogenous POMC in HEK-293T cells and NHEMs. Wt and mutant SASH1, exogenous p53 and exogenous POMC were introduced into HEK-293T cells and NHEMs. At 48-h post-transfection, immunoblot were performed to detect the corresponding protein levels. (C) As identified by IHC analyses, high expression of endogenous p53 was induced by Y551D-SASH1 mutation and more p53-positive epithelial cells were detected in the affected epithelial tissues. Affected epithelial tissues with Y551D SASH1 mutation from pedigree I as well as normal epithelial tissues were fixed and embedded in paraffin for immunohistochemistry detection. Scale bar: 20 μm. The representative positive cells of p53 are donated by red arrows. (D) and (E) Western blot indicated that increased endogenous p53 was induced by SASH1 mutations in HEK-293 cells and NHEMs.

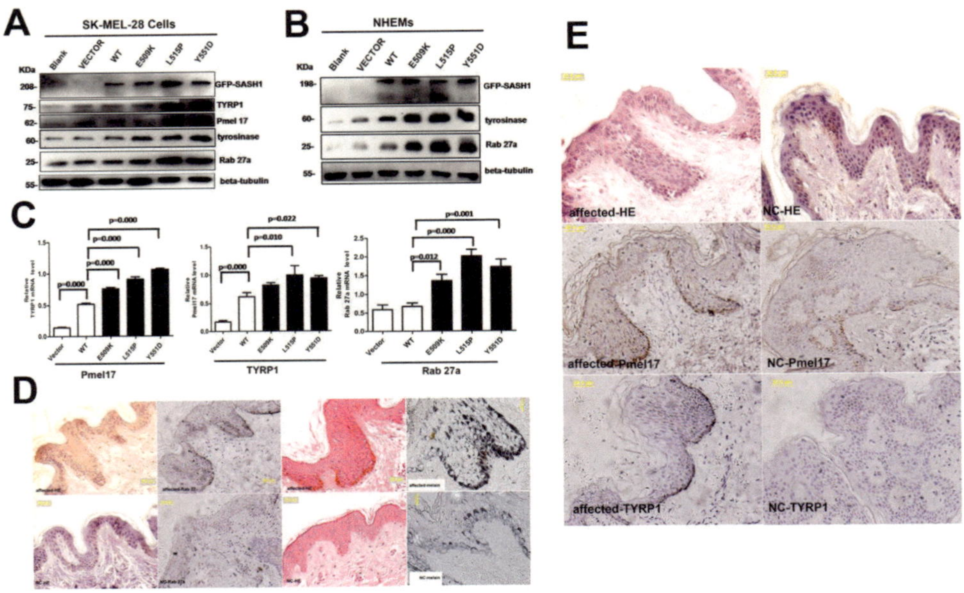

Figure 11.
Increased production of melanogenic components and heterogeneous distribution of melanin in vivo were induced by SASH1 mutations. (A) Up-regulation of melanogenic components including TYRP1, Pmel17, tyrosinase, and Rab 27a were induced by SASH1 mutations in stable SK-MEL-28 cells. (B) Up-regulation of Rab 27a and tyrosinase was also induced by SASH1 mutations. The SASH1 gene (wt and mutant) was introduced into NHEMs and western blot was performed to determine the effect of SASH1 mutations on melanogenic components. (C) As identified by QRT-PCR, up-regulation of Pmel17, TYRP1, and Rab 27a in stable SK-MEL-28 cells was induced by SASH1 mutations (n = 4, mean ± standard error). (D) As identified by immunochemistry detection, heterogeneous distribution of Rab 27a and melanin were observed in the epithelial layers of the affected individuals. (E) Immunochemical analyses indicated that expression of Pmel17 and TYRP1 was heterogeneously distributed in all of the epithelial layers of the epidermal tissues from the DUH-affected individuals. Pmel17, TYRP1, and Rab 27a: 400× magnification, bar = 20 μm; melanin: 1000 magnification. Scale bar: 20 μm. The representative positive cells of Rab 27a, Pmel17, TYRP1, and melanin were indicated by red arrows.

of mutated SASH1 on the protein levels of matrix proteins and transport proteins. Enhanced expression of TYRP1, Rab 27a, Pmel17, and tyrosinase in SK-MEL-28 cells, a pigmented melanoma cell line and NHEMs was significantly induced by SASH1 mutations (**Figure 11A** and **B**). QRT-PCR also indicated that Pmel17, TYRP1, and Rab 27a were up-regulated by mutations of SASH1 in SK-MEL-28 stable cells (**Figure 11C**). Pmel17, TYRP1, and Rab 27a was heterogeneously distributed in the epithelial cells in the tissues of DUH-affected individuals as demonstrated by IHC analyses (**Figure 11D** and **E**). Increased levels of melanogenesis molecules were observed in some hyperpigmentation areas in the affected epithelial layers. In the hyperpigmentation plaques, the superfluous production and secretion of melanin was clearly presented in the basal layers and in the suprabasal layers of the affected epidermal as (**Figure 11D**).

4. Conclusion

Our study reveals that a novel p53-SASH1 reciprocal induction triggers pigmentation of skin through an autoregulatory p53/POMC/α-MSH/Gαs/SASH1 loop. SASH1 mutations enhance SASH1-mediated induction of p53 and POMC. POMC is induced by p53 overexpression and resulted in UV-dependent hyperpigmentation UV-independent pathological hyperpigmentation [1]. Our work indicates that POMC up-regulation is induced by SASH1 mutations, which ultimately results in the pathological hyperpigmentations of affected DUH individuals. These data indicate that SASH1 activation induced by mutations in melanocytes acts as a "UV sensor/effector" for skin pigmentation or SASH1 mutations-mediated up-regulation is the "chief criminal" of pathological hyperpigmentation of DUH, and its underlying mechanistic role is SASH1-p53 reciprocal inducement. Our data indicate that the definitions of the positive feedback p53/POMC/α-MSH/Gαs/SASH1 loop help us to recognize an important linkage between the p53 pathway and MC1R pathway by SASH1.

Recently, a c.1067T>C (p.Leu356Pro) mutation in exon 3 of ABCB6 (ATP-binding cassette subfamily B, member 6) was found in a large five-generation Chinese family with DUH family. Two additional missense mutations, c.508A>G (p.Ser170Gly) in exon 1 and c.1736G>A (p.Gly579Glu) in exon 12 of ABCB6 were found in two out of six patients using sporadic DUH patients [25]. Ac.1663C>A, (p.Gln555Lys) missense mutation in ABCB6 was identified in a Chinese family with typical features of autosomal dominant DUH. Two deletion mutations (g.776 delC, c.459 delC) in ABCB6 were found in an unrelated sporadic affected individual [26]. In addition, missense mutations in ABCB6 were also found in the sporadic affected DUH individuals [27, 28]. Silence of ABCB6 by siRNA destroyed PMEL amyloidogenesis in early melanosomes and resulted in aberrant increase of multilamellar aggregates in pigmented melanosomes. In the retinal pigment epithelium of ABCB6 knockout mice, morphological analysis indicated an obvious decrease of melanosome numbers [29]. All of these sequencing results and functional analyses of causing genes responsible for DUH indicate there exist novel pathogenicity genes and novel gene variations which is responsible for pathogenesis of DUH or there exists novel subtype of DUH.

The transcriptional network of p53-responsive genes produces proteins that interact with a large number of other signal transduction pathways in the cell and a number of positive and negative autoregulatory feedback loops act upon the p53 response [30]. Feedback loops of p53 and p53-responsive genes provide a means to connect the p53 pathway with other signal transduction pathways and coordinate the cellular signals for growth and division [30]. Our findings suggest that SASH1 serves as a "Hinge" to connect p53/POMC/α-MSH pathway with MC1R/Gαs/cAMP/PKA cascade to form an autoregulatory p53/POMC/Gαs/SASH1 circuit to mediate the melanogenesis process [18, 31].

Most recently, SASH1 is showed to involve in autosomal-dominant lentiginous [32] and autosomal-recessive SASH1 variants (c.1849G>A; p.Glu617Lys), which are associated with a new genodermatosis with a pigmentation defects, palmoplantar keratoderma and skin carcinoma and SASH1 is first reported to be predisposed to skin cancer [33]. Dyschromatosis universalis hereditaria (DUH) is a clinically heterogeneous disorder that presents as generalized mottled pigmentation and was first reported by Ichigawa and Hiraga in 1933 [7]. Stuhrmann and colleagues identified the first locus responsible for autosomal-recessive DUH, and this findings is consistent with recent evidence demonstrating that DSH and DUH are genetically distinct disorders [34]. Zhang et al. mapped the causative gene of DSH to 1q11-1q21 and found that a novel mutation of a heterozygous nucleotide A → G at position 2879 in exon 10 of the DSRAD gene is involved in DSH [35]. Subsequent research on dyspigmentation has demonstrated that the pathogenic genetic variant that causes DSH is localized to the DSRAD gene on chromosome 1q [15, 36–41]. Expanding Stuhrmann and Nuber's findings and our own previous work providing photographic evidence of dyschromatosis presenting as large hyperpigmented bodies on DUH-affected individuals [6, 8, 34], we believe that we have discovered the first locus associated with autosomal dominant DUH, identifying SASH1 as the causative gene of autosomal dominant DUH.

Our findings first identify the pathological gene of DUH and reveal the pathological mechanism of hyperpigmentation patches of DUH-affected individuals. In addition, our work will enrich the crosstalk of p53 pathway with other transduction pathways in cells and give a new definition of the p53-responsive genes and their associations, which will perfect the p53 programmed responses to stress and pathologic conditions.

Acknowledgements

We thank Central Laboratory at Yongchuan Hospital, Chongqing Medical University and Clinical Research Center, the Affiliated Hospital, Guizhou Medical University for housing the experiments. This work was supported by the Shanghai Municipal Commission of Science and Technology Program (09DJ1400601), the 973 Program (2010CB529600, 2007CB947300), the Yongchuan Hospital Project, Chongqing Medical University (YJYJ201347), Chongqing Education Commission Project (KJ1400201), and Guizhou's Introduction Project of Million Talents.

Conflict of interest

No conflict between the authors.

Notes/thanks/other declarations

The chapter text was mainly referred to our article entitled as "A Novel P53/POMC/Gαs/SASH1 Auto-regulatory Feedback Loop Activates Mutated SASH1 to Cause Pathologic Hyper-pigmentation" (Journal of Cellular and Molecular Medicine 2017, 21(4):802-815) which we published in journal of cellular and molecular medicine in April, 2017. In this chapter, we rewrite the chapter text according to the suggestions of reviewers.

The chapter figures were taken from the figures and supplementary figures of our article entitled as "A Novel P53/POMC/Gαs/SASH1 Auto-regulatory Feedback Loop Activates Mutated SASH1 to Cause Pathologic Hyper-pigmentation". The chapter tables were taken from the supplementary tables of our published article.

Declarations

Appendices and nomenclature

ABCB6	ATP-binding cassette subfamily B, member 6
ACTH	adrenocorticotropic
α-MSH	α-melanocyte stimulating hormone

cAMP	cyclic adenosine monophosphate
CDS	coding sequence
CHX	cycloheximide
DSRAD	double-stranded RNA-specific adenosine deaminase
DSH	dyschromatosis symmetrica hereditaria
DUH	dyschromatosis universalis hereditaria
EMSA	electrophoretic mobility shift assay
GAPDH	glyceraldehyde phosphate dehydrogenase
Gαs	guanine nucleotide-binding protein subunit-alpha isoforms short
GPCR	G-protein-coupled receptor
GS	Griscelli syndrome
IF	immunofluorescence
IHC	immunohistochemical
LC-MS/MS	liquid chromatograph-mass spectrometer
LOD	log odds
LSD	least-significant-difference
MC1R	melanocortin-1-receptor
Mitf	microphthalmia-associated transcription factor
NHEMs	normal human epithelial melanocytes
PEI	polyethyleneimine
POMC	pro-opiomelanocortin
QRT-PCR	quantitative reverse transcriptase polymerase chain reaction
RT-PCR	reverse transcriptase polymerase chain reaction
SASH1	SAM and SH3 domain containing 1
SED	standard erythema dose
TYRP1	tyrosinase protein1
UV	ultraviolet

Author details

Ding'an Zhou[1]*, Jiawei Zeng[2], Xing Zeng[3], Yadong Li[4], Zhixiong Wu[4], Xin Wan[4], Pingshen Hu[1] and Xiaodong Su[1]

1 Clinical Research Center, The Affiliated Hospital, Guizhou Medical University, Guiyang, Guizhou, China

2 Department of Clinical Laboratory, Mianyang Central Hospital, Mianyang, Sichuan, China

3 The Second Affiliaed Hospital, Chongqing Medical University, Chongqing, China

4 Clinical Medical College, Guizhou Medical University, Guiyang, Guizhou, China

*Address all correspondence to: 460318918@qq.com

References

[1] Cui R et al. Central role of p53 in the suntan response and pathologic hyperpigmentation. Cell. 2007;**128**(5):853-864

[2] Roberts DW et al. Quantitative analysis of MC1R gene expression in human skin cell cultures. Pigment Cell Research. 2006;**19**(1):76-89

[3] Rees JL. The genetics of sun sensitivity in humans. American Journal of Human Genetics. 2004;**5**(75):739-751

[4] Jiang Y et al. Regulation of G-protein signaling by RKTG via sequestration of the G betagamma subunit to the Golgi apparatus. Molecular and Cellular Biology. 2010;**30**(1):78-90

[5] Ananthaswamy HN et al. Sunlight and skin cancer: Inhibition of p53 mutations in UV-irradiated mouse skin by sunscreens. Nature Medicine. 1997;**3**(5):510-514

[6] Nuber UA et al. Dyschromatosis universalis hereditaria: Familial case and ultrastructural skin investigation. American Journal of Medical Genetics. Part A. 2004;**125A**(3):261-266

[7] Ichigawa TH. A previously undescribed anomaly of pigmentation dyschromatosis universalis hereditaria. The Japanese Journal of Dermatology. 1933;**34**:360-364

[8] Xing QH, Wang MT, Chen XD, Feng GY, Ji HY, Yang JD, et al. A gene locus responsible for dyschromatosis symmetrica hereditaria (DSH) maps to chromosome 6q24.2-q25.2. American Journal of Human Genetics. 2003;**73**(2):377-382

[9] Zhong WL et al. Novel mutations in SASH1 associated with dyschromatosis universalis hereditaria. Indian Journal of Dermatology, Venereology and Leprology. 2018. DOI: 10.4103/ijdvl.IJDVL_360_17

[10] Zeller C, Hinzmann B, Seitz S, Prokoph H, Burkhard-Goettges E, et al. SASH1: A candidate tumor suppressor gene on chromosome 6q24.3 is downregulated in breast cancer. Oncogene. 2003;**22**:2972-2983

[11] Rimkus C et al. Prognostic significance of downregulated expression of the candidate tumour suppressor gene SASH1 in colon cancer. British Journal of Cancer. 2006;**95**(10):1419-1423

[12] Martini M et al. The candidate tumor suppressor SASH1 interacts with the actin cytoskeleton and stimulates cell-matrix adhesion. The International Journal of Biochemistry & Cell Biology. 2011;**43**(11):1630-1640

[13] Zhou D, Wei Z, Deng S, Wang T, Zai M, Wang H, et al. SASH1 regulates melanocyte transepithelial migration through a novel Galphas-SASH1-IQGAP1-E-cadherin dependent pathway. Cellular Signalling. 2013;**25**:1526-1538

[14] Satoh Y, Yoshida M. Clinical and photobiological differences between dyschromatosis symmetrica hereditaria and xeroderma pigmentosum. The Journal of Dermatology. 1980;7:317-322

[15] Mataraza JM et al. IQGAP1 promotes cell motility and invasion. The Journal of Biological Chemistry. 2003;**278**(42):41237-41245

[16] Shi X, Sun X, Liu M, Li D, Aneja R, Zhou J. CEP70 protein interacts with gamma-tubulin to localize at the centrosome and is critical for mitotic spindle assembly. The Journal of Biological Chemistry. 2011;**286**(38):33401-33408

[17] Palavalli LH, Prickett T, Wunderlich JR, Wei X, Burrell AS, Porter-Gill P, et al. Analysis of the matrix metalloproteinase family reveals that MMP8 is often mutated in melanoma. Nature Genetics. 2009;**41**:518-520

[18] Zhou D'a, Wei Z, Kuang Z, Luo H, Ma J, Zeng X, et al. A novel P53/POMC/Gαs/SASH1 auto-regulatory feedback loop activates mutated SASH1 to cause pathologic hyper-pigmentation. Journal of Cellular and Molecular Medicine. 2017;**21**(4):802-815

[19] Zhou X, Ting-Ting L, Feng X, Hsiang E, Xiong Y, Guan K-L, et al. Targeted polyubiquitylation of RASSF1C by the mule and SCFβ-TrCP ligases in response to DNA damage. The Biochemical Journal. 2012;**441**:227-236

[20] Schiaffino M. Signaling pathways in melanosome biogenesis and pathology. The International Journal of Biochemistry & Cell Biology. 2010;**42**(7):1094-1104

[21] Herraiz C et al. Signaling from the human melanocortin 1 receptor to ERK1 and ERK2 mitogen-activated protein kinases involves transactivation of cKIT. Molecular Endocrinology. 2011;**25**(1):138-156

[22] Tsatmali M, Ancans J, Yukitake J, Thody AJ. Skin POMC peptides: Their actions at the human MC-1 receptor and roles in the tanning response. Pigment Cell Research. 2000;**13**(Suppl 8):125-129

[23] D'Orazio JA, Nobuhisa T, Cui R, Arya M, Spry M, Wakamatsu K, et al. Topical drug rescue strategy and skin protection based on the role of Mc1r in UV-induced tanning. Nature. 2006;**443**:340-344

[24] Diffey BL, Jansen CT, Urbach F, Wulf HC. The standard erythema dose: A new photobiological concept. Photodermatology, Photoimmunology & Photomedicine. 1997;**13**(1-2):64-66

[25] Zhang C et al. Mutations in ABCB6 cause dyschromatosis universalis hereditaria. The Journal of Investigative Dermatology. 2013;**133**(9):2221-2228

[26] Cui YX et al. Novel mutations of ABCB6 associated with autosomal dominant dyschromatosis universalis hereditaria. PLoS One. 2013;**8**(11):e79808

[27] Liu H, Li Y, Kwok HK, et al. Genome-wide linkage, exome sequencing and functional analyses identify ABCB6 as the pathogenic gene of dyschromatosis universalis hereditaria. PLoS One. 2014;**9**(2):e87250

[28] Lu C, Liu J, Liu F, Liu Y, Ma D, Zhang X. Novel missense mutations of ABCB6 in two chinese families with dyschromatosis universalis hereditaria. Journal of Dermatological Science. 2014;**76**(3):255-258

[29] Bergam P et al. ABCB6 resides in melanosomes and regulates early steps of melanogenesis required for PMEL amyloid matrix formation. Journal of Molecular Biology. 2018;**S0022-2836**(18):30662-30664

[30] Harris SL, Levine AJ. The p53 pathway: Positive and negative feedback loops. Oncogene. 2005;**24**(17):2899-2908

[31] Zhou D et al. p53 regulates ERK1/2/CREB cascade via a novel SASH1/MAP2K2 crosstalk to induce hyperpigmentation. Journal of Cellular and Molecular Medicine. 2017;**21**(10):2465-2480

[32] Shellman YG et al. SASH1 is involved in an autosomal dominant lentiginous phenotype. The Journal of Investigative Dermatology. 2015;**135**(12):3192-3194

[33] Courcet JB et al. Autosomal-recessive SASH1 variants associated

with a new genodermatosis with pigmentation defects, palmoplantar keratoderma and skin carcinoma. European Journal of Human Genetics. 2015;**23**(7):957-962

[34] Stuhrmann M et al. Dyschromatosis universalis hereditaria: Evidence for autosomal recessive inheritance and identification of a new locus on chromosome 12q21-q23. Clinical Genetics. 2008;**73**(6):566-572

[35] Zhang XJ, Gao M, Li M, Li M, Li CR, et al. Identification of a locus for dyschromatosis symmetrica hereditaria at chromosome 1q11-1q21. The Journal of Investigative Dermatology. 2003;**120**:776-780

[36] Li W, Li H, Bocking AD, Challis JR. Tumor necrosis factor stimulates matrix metalloproteinase 9 secretion from cultured human chorionic trophoblast cells through TNF receptor 1 signaling to IKBKB-NFKB and MAPK1/3 pathway. Biology of Reproduction. 2010;**83**(3):484-487

[37] Li M et al. Mutational spectrum of the ADAR1 gene in dyschromatosis symmetrica hereditaria. Archives of Dermatological Research. 2010;**302**(6):469-476

[38] Liu Y et al. Two novel frameshift mutations of the DSRAD gene in Chinese pedigrees with dyschromatosis symmetrica hereditaria. International Journal of Dermatology. 2012;**51**(8):920-922

[39] Ren JW et al. Novel frameshift mutation of the DSRAD gene in a Chinese family with dyschromatosis symmetrica hereditaria. Journal of the European Academy of Dermatology and Venereology. 2008;**22**(11):1375-1376

[40] Wang XP et al. Four novel and two recurrent mutations of the ADAR1 gene in Chinese patients with dyschromatosis symmetrica hereditaria. Journal of Dermatological Science. 2010;**58**(3):217-218

[41] Xing Q et al. Identification of a novel ADAR mutation in a Chinese family with dyschromatosis symmetrica hereditaria (DSH). Archives of Dermatological Research. 2005;**297**(3):139-142

Chapter 7

Molecular Diagnosis of Invasive Aspergillosis

María del Rocío Reyes-Montes, Esperanza Duarte-Escalante, María Guadalupe Frías-De-León, Erick Obed Martínez-Herrera and Gustavo Acosta-Altamirano

Abstract

Invasive aspergillosis (IA) is a disease that is difficult to manage and is associated with a significantly high morbidity and mortality, caused by different species of the genus *Aspergillus*, and closely related to immunocompromised patients; thus, it is important to understand the distribution and molecular epidemiology of the species causing this disease. Even though *Aspergillus fumigatus sensu stricto* is the most common species that cause IA, in recent years, there has been an increase in the number of species in the different sections which makes the diagnosis of this invasive fungal disease a great challenge. Conventional tests for the diagnosis of IA present limitations in sensitivity and specificity, while molecular tests have the potential to improve diagnosis by offering a more sensitive and rapid identification, but they are not yet standardized for reliable use in clinic. Nevertheless, there are some tests for the presumptive diagnosis of aspergillosis which, although are not specific for the identification of species, have been decisive in the case of IA. Among these are the Galactomannan test (GM), the Beta-D-glucan assay and volatile organic compounds (VOCs) testing. In this chapter, the recent advances and challenges in the molecular diagnosis of IA are revised.

Keywords: molecular diagnosis, invasive aspergillosis, PCR, epidemiology, molecular markers

1. Introduction

Aspergillosis is defined as tissue damage caused by fungi of the genus *Aspergillus* [1], which belongs to the class Ascomycetes. This genus consists of 8000 opportunistic and saprobic fungal species that have been reclassified as 250 species in nine primary sections: *Flavi*, *Fumigati*, *Nigri*, *Udagawae*, *Cricumdati*, *Versicolor*, *Usti*, *Terrei* and *Emericella* [2, 3]. Members of the genus *Aspergillus* are filamentous fungi that are ubiquitous in the environment [4, 5] and primarily develop close to decomposing plants, organic waste and soil, where they produce large numbers of conidia that are disseminated through the air [6].

Aspergillus is a cosmopolitan fungus that primarily infects immunocompromised hosts and individuals with the underlying lung disease. Some *Aspergillus* species are capable of causing a wide variety of diseases in humans and animals, collectively known as aspergillosis, which have increased significantly in recent years [7]. Approximately, 45 species have been reported to cause disease in humans [2, 8].

Aspergillosis can cause several clinical symptoms in humans, especially in immunocompromised individuals [8]. *A. fumigatus sensu stricto* is the most important opportunistic human pathogen, producing thousands of airborne conidia, which, due to their small size (2.5–3.5 μm in diameter), can be disseminated great distances by atmospheric disturbances such as wind convection currents and can survive under a wide range of environmental conditions. These airborne conidia can eventually be inhaled into the lungs of humans and animals, where an efficient innate immune response is required to prevent infection [8]. The primary diseases associated with this pathogen are allergic bronchopulmonary aspergillosis, chronic pulmonary aspergillosis, and IA, the latter of which is difficult to manage, leading to a significantly high morbidity and mortality. IA is the most serious disease caused by this fungus since it involves the invasion of fungal hyphae into tissue, and in some cases, haematogenously spread to other organs, particularly the brain [8]. The primary site of IA infection is the lungs. Infections of the skin and cornea may also occur, but fungal colonization of these sites is much less frequent. IA is rare in healthy individuals and almost exclusively affects patients with compromised immune systems [8–13]. The prognosis for these patients is often poor due to the scarcity of effective treatments combined with the already compromised state of health of these individuals, with mortality rates ranging between 30 and ≥90% depending on the immune status of the patient [14]. To reduce mortality in patients with IA, von Eiff et al. [15] administered empirical antifungal therapies to patients with severe neutropaenia (absolute granulocytosis of ≤500/μl) and observed a 90–41% reduction in mortality. However, this non-specific strategy has caused an increase in resistance to antifungals in patients undergoing long-term treatment [16]. In addition, some species belonging to the complex *A. fumigatus sensu latu* (*Fumigati* section) have been recognized as occasional causes of IA [17–20], and other IA-causing species belong to different *Aspergillus* sections or complexes (**Table 1**).

Country	*Total	Rate/100,000	Reference
Algeria	2865	7.1	[21]
Bangladesh	5166	3.2	[22]
Belgium	675	6.08	[23]
Brazil	8664	4.47	[24]
Burkina Faso	54[a]	0.3	[25]
Cameroon	1175	5.3	[26]
Canada	566	1.59	[27]
Chile	296	1.7	[28]
Colombia	2820	5.7	[29]
Czech Republic	297	2.8	[30]
Ecuador	594	1.3	[31]
Egypt	9001	10.7	[32]
France	1185	1.8	[33]
Greece	1125	10.4	[34]
Guatemala	671	4.4	[35]
Hungary	319[b]	3.2	[36]
Jordan	84	1.34	[37]
Kazakhstan	511	2.8	[38]
Korea	2150	4.48	[39]
Malaysia	1018	3.3	[40]

Country	*Total	Rate/100,000	Reference
Mexico	20[c]	0.017	[41]
	4[d]	0.0036	
Nepal	1119	4	[42]
Norway	278	5.3	[43]
Pakistan	10949	5.9	[44]
Peru	1621	5	[45]
Portugal	240	2.3	[46]
Philippines	3085	3	[47]
Qatar	11	0.6	[48]
Romania	1524	7.7	[49]
Russia	3238	2.27	[50]
Spain	4318	9.19	[51]
Thailand	941	1.4	[52]
Tanzania	20	0.05	[53]
Trinidad and Tobago	8	0.6	[54]
Ukraine	1233	2.7	[55]
Uruguay	773	22.4	[56]
United Kingdom	2001-2912[e]	4.59-4.6	[57]
Vietnam	14523	15.9	[58]

[a]*Haematological malignancies*

[b]*Chronic obstructive pulmonary disease.*

[c]*Renal, heart and liver transplant recipients.*

[d]*Allogenic- hematopoietic stem cell transplant.*

[e]*All risk groups except critical care patients.*

**Total expected number of cases.*

Table 1.
Estimated burden of invasive aspergillosis.

2. Taxonomy of the genus *Aspergillus*

The genus *Aspergillus* belongs to the order Eurotiales and includes over 344 species, more than 40 of which are aetiological agents of opportunistic human infections, although some of them do so only occasionally [59].

Aspergillus is traditionally classified based on morphological characteristics, such as the size and arrangement of the aspergillary heads, the color of the conidia, the growth rate in different media and physiological characteristics. According to these morphological characteristics, Raper and Fennell [60] divided the genus *Aspergillus* into 18 groups. However, because this classification did not have any status in the nomenclature, Gams et al. [61] introduced the use of *Aspergillus* subgenera and sections. These studies showed that the groups organized by Raper and Fennell [60], which were based on phenotypic characteristics, largely coincide with the current classifications. However, because morphological variations in several sections resulted in controversial taxonomic groups, polyphasic identification was used, which involves the morphological, physiological, molecular and ecological characterization of a species [3]. Peterson [62] established the acceptance of five subgenera (*Aspergillus*, *Circumdati*, *Fumigati*, *Nidulantes* and *Ornati*) with 16 sections from a phylogenetic analysis rDNA region sequences. By contrast, Samson and Varga [63], based on phylogenetic analysis using multilocus sequence typing (using calmodulin, RNA polymerase 2 and the rRNA gene), subdivided *Aspergillus* into eight subgenres: the subgenus *Aspergillus*, with the sections *Aspergillus* and *Restricti*; the subgenus *Fumigati*, with the sections *Fumigati*, *Clavati* and *Cervini*; the subgenus *Circumdati*, with the sections *Circumdati*, *Nigri*, *Flavi* and *Cremei*; the subgenus *Candidi*, with the section *Candidi*; the subgenus *Terrei*, with the sections *Terrei* and *Flavipedes*; the subgenus *Nidulantes*, with the sections *Nidulantes*, *Usti* and *Sparsi*; the subgenus *Warcupi*, with the sections *Warcupi* and *Zonati*; and the subgenus *Ornati*, with the section *Ornati*. Later, Varga et al. [64], based on multilocus sequence typing (using β-tubulin, calmodulin and the intergenic spacing regions [ITS] region), added the subgenus *Nidulantes* to the *Aenei* section.

Subsequently, based on different studies [62, 64–66], a new classification was proposed for the genus *Aspergillus* that included four subgenres and 19 sections in which the subgenera *Ornati* and *Warcupi* were transferred to other genera, since they did not belong to the genus *Aspergillus*. Similarly, the *Cremei* section, which had been classified into the subgenus *Aspergillus* [62], was reclassified into the subgenus *Circumdati* by Houbraken and Samson [66], resulting in the following classification: the subgenus *Aspergillus*, with the sections *Aspergillus* (teleomorph *Eurotium*) and *Restricti* (teleomorph *Eurotium*); the subgenus *Circumdati*, with the sections *Candidi*, *Circumdati* (teleomorph *Neopetromyces*), *Flavi* (*Petromyces*), *Flavipedes* (*Fennellia*), *Nigri* and *Terrei*; the subgenus *Fumigati*, with the sections *Cervini*, *Clavati* (teleomorph *Neocarpenteles*, *Dichotomomyces*) and *Fumigati* (*Neosartorya*); and the subgenus *Nidulantes*, with sections *Aeni* (teleomorph *Emericella*), *Bispori*, *Cremei* (teleomorph *Chaetosartorya*), *Nidulantes* (teleomorph *Emericella*), *Ochraceorosei*, *Silvati*, *Sparsi* and *Usti* (teleomorph *Emericella*). Finally, the current classification consists of four subgenera (*Aspergillus*, *Circumdati*, *Fumigati* and *Nidulantes*) and 20 sections [67, 68], with 339 correctly identified species [59]. There are currently 45 species of *Aspergillus* described as human pathogens [2], although the number of clinically relevant fungal species has been steadily increasing in recent years and is likely to increase further in the future (**Table 1**).

Colonies of the genus *Aspergillus* are typically fast growing and can exhibit a range of colors, including white, yellow, yellow-brown, brown and black or exhibit shades of greenish-gray or blue-green. *Aspergillus* species are characterized by the production of specialized structure, called conidiophores, which in some cases can be dramatically different [59]. Although it is a unicellular structure, the conidiophore has three distinct parts: the vesicle (swollen apical end), the stipe (cylindrical section located below the vesicle) and the foot cell (final section, sometimes separated by a septum that joins the conidiophore with the mycelium). The vesicle is partially or entirely covered by an enclosure of phialides (uniseriate). In many species, other cells called metulae are located between the vesicle and the phialides that support a small number of compact phialides (biseriate). The set forms what is called the aspergilar head, which can be strictly uniseriate, biseriate or mixed. The conidia are unicellular, round, oval, elliptical, smooth or rough, hyaline or pigmented, with thick or thin walls, produced in long chains that can be divergent (radiated) or aggregated in compact (columnar) or lax columns (that tend to open). Few species produce other types of conidia besides the phialides and metulae on the vesicle, and most develop directly on the vegetative hyphae as round or ovoid forms called aleuroconidia. Some species can produce Hülle cells, which can be solitary or envelop the cleistothecia form of one of the associated teleomorphs. The types of known ascomata present great morphological variation, ranging from those surrounded by loose hyphae with a smooth pseudoparenchymatous appearance to those exhibiting compact sclerotium structures, with variations in size, ornamentation and type of ascospore surfaces also observed [69].

The classification of the genus *Aspergillus* into subgenres and sections is made based on four fundamental characteristics: the presence of a teleomorph, the presence or absence of metulae, the disposition of the metulae or phialides on the vesicle and the colouration of the colonies [69].

3. Epidemiology

The number of IA cases has increased in recent years. Recent data provided by Bongomin et al. [70] estimate approximately 250,000 cases that occur worldwide each year, with an associated increase in morbidity and mortality rates. This increase is of great importance for health-care systems, since the epidemiological surveillance systems in several countries are inefficient. When combined with the difficulties in diagnosing IA, the resulting delay in the application of timely treatment can lead to the death of patients (**Table 2**). Given the importance of this disease, several research groups have developed 'The *Aspergillus* guide', which includes diagnostic and therapeutic guidance, focusing on life-threatening diseases caused by *Aspergillus* spp., primarily in Europe [71].

The major risk factors for developing IA are neutropaenia, allogeneic transplantation of hematopoietic stem cells or solid organ transplantation (particularly lung), haematologic malignancy and chemotherapy with cytotoxic cancer. Patients with chronic granulomatous disease and advanced AIDS also have a high risk of developing IA, as do patients receiving treatment with chronic steroid therapies and tumor necrosis factor as well as those with long-term chronic

Species	Identification method	Reference
A. ustus (100)	Culture and molecular typing (RAPD)	[72]
Emericella quadrilineata (4)	Sequence-based analysis [ITS region, β-tubulin (*benA*) and calmodulin (*caM*)]	[73]
A. viridinutans	Sequence-based analysis (β -tubulin and rodlet A gene)	[74]
A. udagawae	Sequence-based analysis [β-tubulin (*benA*)]	[75]
A. fumigatus (32) *A. lentulus* (4) *A. calidoustus* (2) *A. tubingensis* (1) *A. sydowii* (1) *A. flavus* (1) *A. terreus* (1) *E. rugulosa* (1)	Sequence-based analysis [ITS region, β-tubulin (*benA*) and calmodulin (*caM*)]	[76]
Section *Usti*: *A. calidoustus* Section *Fumigati*: *A. novofumigatus* *A. viridinutans*	Sequence-based analysis [β-tubulin (*benA*)]	[77]
E. nidulans var. *echinulata*	Sequence-based analysis [ITS region, β-tubulin (*benA*) and calmodulin (*caM*)]	[78]
A. lentulus	Sequence-based analysis [ITS region, β-tubulin (*benA*) and calmodulin (*caM*)]	[79]
A. lentulus	Culture Thermotolerence Sequence-based analysis [ITS region, β-tubulin (*benA*)	[80]

Species	Identification method	Reference
Culture: *A. fumigatus* (72.7) *A.flavus* (27.3) Nested PCR: *A. fumigatus* (30) *A.flavus* (20) Real-time PCR: *A. fumigatus* (6.25) *A. flavus* (12.5)	Culture (standard morphological procedures) Nested PCR Real-time PCR	[81]
Section *Circumdati*: *A. pallidofulvus* (6.66) *A. ochraceus* (2.22) Section *Flavi*: *A. tamari* (8.88) Section *Terrei*: *A. niveus* (6.66) Section *Versicolor*: *A. sydowii* (8.88) Section *Aspergillus*: *A. montevidensis* (6.66) Section *Nigri*: *A. brunneoviolaceus* (4.44)	Culture (standard morphological procedures) Sequence-based analysis [β-tubulin (*benA*) and calmodulin (*caM*)] MALDI-TOF MS	[82]
A. fumigatus (28.8) *A. flavus* (7.2)	Sequence-based analysis: [ITS region, β-tubulin (*benA*)]	[83]
A. flavus complex (15) *A. tubingensis* (3) *A. fumigatus* (2)	Sequence-based analysis: (region, β-tubulin (*benA*)	[84]
Section *Nidulantes*: *A. sublatus* (1)	Scanning electron microscopy of ascospores Sequencing of calmodulin gene	[85]

Table 2.
Species associated with invasive aspergillosis (IA).

diseases and other conditions, such as diabetes mellitus, rheumatological conditions, liver disease and chronic obstructive disease [8–13, 86–88]. It is also important to know the distribution and molecular epidemiology of *Aspergillus* species obtained from clinical and environmental sources in different geographical regions of the world, since different *Aspergillus* species can cause IA. In addition, although the *Fumigati* section (complex *A. fumigatus*) has been reported as the most frequent cause of IA, data suggest that IA can be caused by other species in immunocompromised hosts (**Table 2**), especially *A. niger*, *A. terreus* and species of the complex *A. flavus.* Therefore, the precise identification of *Aspergillus* species isolated from patients is of great importance for the selection of an effective antifungal therapy [84].

Because *Aspergillus* species are widely distributed fungi in the environment, and their conidia are dispersed primarily through air currents, their relationship with hosts in hospital environments is of great relevance. Standards have been established for hospital environments for adults and immunosuppressed children who require special attention. Ullmann et al. [71] recommend that patients should be separated from areas under construction or renovation and from potted plants and flowers in patient rooms and living quarters. In addition, they recommend placing patients in special rooms with positive air pressure and HEPA filters or laminar air flow and rooms with filters for water supplies, especially in showers.

4. Molecular diagnosis

Accurate and early diagnosis of an active *Aspergillus* infection is necessary to initiate effective antifungal therapy, particularly in critically ill patients [89]. The IA diagnosis is made based on the criteria defined for proven, probable or possible infection, implemented by the European Organization for Cancer Research and Treatment/Study Group on Mycoses (EORTC/MSG). These criteria depend on the clinical manifestations, host and fungal factors as well as the results of traditional laboratory methods (histopathology and culture) [90, 91]. However, the diagnosis of IA continues to be a challenge, since histopathology and cultures have limitations in their sensitivity and specificity as well as in the time required to obtain the results, which leads to significant delays in the initiation of treatment [89]. To achieve an accurate diagnosis and timely and effective treatment, molecular tests have been developed that overcome the limitations of conventional methods and can reduce the mortality rate associated with IA [92]. Thus, Samson et al. [59] suggested using a polyphasic approach (morphological characterization, physiological tests, ecological data, extrolite analysis and DNA sequencing) as a gold standard for the identification of *Aspergillus* species. However, because these methodologies are time and labour intensive, they are not practical in most clinical laboratories. Therefore, different PCR modalities have been developed through the use of specific molecular markers for the detection of *Aspergillus* species of medical importance to quickly identify the aetiological agent of aspergillosis [93, 94].

Molecular tests for the diagnosis of IA have been developed for both home and commercial use to directly detect and identify *Aspergillus* spp. in different clinical specimens, including the following: whole blood, serum, plasma, bronchoalveolar lavage (BAL), sputum, bronchial aspirate, tissue, pleural effusion and cerebrospinal fluid (CSF), allowing for the PCR amplification of fungal DNA via nested, multiplex and real-time PCR [89, 93, 95–105] and with minor frequency via isothermal amplification (loop-mediated isothermal amplification, LAMP) [106]. The efficiency of these tests is variable and depends on many factors, including the DNA extraction method, clinical sample type, type of PCR, amplification target and detection method. The lack of standardization of these technical problems represents the most important barrier for the widespread application of PCR as a diagnostic modality for the diagnosis of IA [107].

4.1 DNA extraction methods

The quality and quantity of DNA available for amplification depends on the extraction method used. Home, commercial and automated methods for DNA extraction are available. In the home, enzymatic, chemical or physical agents are used to break the cell wall, while sodium dodecyl sulphate, beta-mercaptoethanol and ethylenediaminetetraacetic acid are used to lyse the membrane. The elimination of proteins and purification of the DNA is performed by an extraction with phenol-chloroform, after which the purified DNA is precipitated with alcohol [91]. DNA can also be extracted using commercial methods or kits, such as the Qiagen QIAmp Tissue Kit (Hilden, Germany), but they are disadvantageous in that the efficiency of fungal DNA extraction can vary considerably between different commercial brands. Furthermore, for both the home-based and commercial methods, contamination of the extraction systems and reagents has been reported, which contributes to variations in the sensitivity and specificity of the tests. The use of automated methods, such as MagNA Pure LC (Roche Diagnostics, Basel, Switzerland), is a viable and high-performance option for DNA extraction but may be cost-prohibitive in many in-hospital laboratories with limited resources [91, 108].

4.2 Types of clinical samples

The direct detection of *Aspergillus* DNA has primarily been performed in whole blood, serum, plasma, and BAL, and occasionally in other specimens, such as sputum, bronchial aspirate, tissue, pleural effusion, peritoneal fluid and cerebrospinal fluid. It has been observed that the interpretation of results is easier when sterile samples are used than when non-sterile samples are used, such as BAL, since the ubiquity of the fungus can promote its presence in the upper respiratory tract, making colonization, invasion or contamination difficult to determine. To date, the most appropriate clinical sample for the diagnosis of IA has not been defined. However, the use of total blood and its fractions (serum or plasma) are the most widely used, the ease at which both blood can be obtained and the interpretation of results. The optimal blood fraction for the detection of *Aspergillus* DNA is unknown. Studies have been conducted to identify the ideal hematological sample to detect *Aspergillus* DNA, taking into account that the processing of whole blood and plasma requires the use of an anticoagulant, such as EDTA, which could inhibit PCR. It has been observed that the greatest inhibition occurs when using heparin or sodium citrate as an anticoagulant [109]. Some authors consider that from a practical point of view, serum is the best hematological sample for the detection of *Aspergillus* DNA, since it is easier to process and allows for antigens to be detected at the same time [98]. Other authors have shown that the sensitivity of PCR in serum and blood is similar, since the levels of circulating *A. fumigatus* DNA are between 100 fg/ml and 1 ng/ml, both in serum and in whole blood. They also report that the sensitivity may increase when a combination of serum and blood is used, which can perhaps be explained by the fact that the performance of DNA detection improves when large volumes of samples are used, rather than by the combination of the two *per se* [95, 98, 109, 110]. Meanwhile, Springer et al. [111] reported that PCR performed using plasma showed a higher sensitivity (91%) than for serum (80%) and whole blood (55%). These observations contrast with previous reports, possibly due to differences in sample processing [107].

BAL has also been used to detect *Aspergillus* by PCR, showing promising and contradictory results. In clinical studies, the general ranges of sensitivity and specificity varied widely, from 73 to 100% and 80 to 100%, respectively, depending on the characteristics of the trial and the type of patients evaluated. The majority of PCR studies using BAL have been performed using patients with haemato-oncological disorders. While some studies demonstrated a high specificity [100], others have shown greater sensitivity [112] or high sensitivity and specificity [99].

Other types of samples that have been used are lung or other deep-tissue biopsies, CSF and pleural effusion, which provided acceptable values of sensitivity and specificity [102, 103, 113]. However, its applicability in routine laboratory diagnostics is limited by the difficulty in obtaining this type of specimen.

4.3 Types of PCR

Nested PCR has been successfully used to detect *Aspergillus* spp. [102]. However, this technique is not the most suitable format, since although the process allows for great sensitivity, there is also the possibility of contamination and the generation of false-positive results. Multiplex PCR has also been used to detect *Aspergillus* spp., particularly in BAL samples. However, despite its adequate specificity and sensitivity (0.01 ng of DNA) [93], the detection method required (electrophoresis in an agarose gel) limits its application as a diagnostic tool, since results in this setting are expected to be obtained as quickly as possible. The primary limitation of these end-point PCR formats is the inability to differentiate between colonization or active infection. In 2006, the European *Aspergillus* PCR Initiative (EAPCRI) sought proposals for a technical consensus. This consensus was possible thanks to the generalization of real-time quantitative PCR (qPCR). This technique drastically reduces the risk of contamination, allows quantitative management of the amplification reaction (quantification of the fungal load), differentiates between several pathogenic species when multiple probes are used in a single assay and allows for the result to be known; at the same time, the amplification is carried out, which helps establish an effective treatment in a timely manner [114]. Therefore, most of the PCR assays that have been developed for the diagnosis of IA are in a real-time format, primarily using hydrolysis probes (TaqMan) directed at the 18S, 28S and ITS regions [91].

Another alternative to detect *Aspergillus* is LAMP. This technique offers several potential advantages over PCR, among which is a more efficient amplification, the possibility of evaluating cell viability, a reduction in the contamination of the genetic material to be amplified and the possibility of using RNA as a target instead of DNA. The latter advantage allows for a greater sensitivity since highly expressed genes produce thousands of transcripts within a cell [106, 107].

4.4 Molecular marker or amplification targets

The selection of a molecular marker or a target region for PCR amplification to diagnose IA is of great importance, since it must favor a sensitive and specific amplification. The most commonly used markers to achieve high sensitivity in detecting *Aspergillus* spp., both in home and in commercial PCR assays, are multicopy genes, such as the ribosomal 18S and 28S genes, for which hundreds of copies are present in the genomes of *Aspergillus* spp. [89, 95, 97, 99, 102–105]. The disadvantage of using multicopy genes is that they involve highly conserved sequences in fungi, leading to a limited specificity (panfungal PCR). This type of amplification target does not always allow the differentiation of phylogenetically related species, such as species that are within the same section, which can often be important. For example, in the *Fumigati* section, although *A. fumigatus*, *A. lentulus* and *N. udagawe* are closely related, the latter two species are more resistant to antifungals than *A. fumigatus*. This PCR method may not even be able to distinguish between phylogenetically related genera (*Aspergillus*, *Penicillium* and *Paecilomyces* spp.), representing a problem in the selection of antifungal treatments [99]. Strategies have been designed to overcome the lack of specificity of multicopy genes, such as hybridization with species-specific probes [112]. Similarly, the variable regions of the rRNA gene, the ITS regions and D1/D2 are used [98, 100, 115]. ITS sequences may lack sufficient variation for the resolution of some *Aspergillus* species, and a bar code or a secondary identification marker is typically needed to identify an isolate at the species level. Based on these observations, the use of sequencing gene fragments, such as α-tubulin (*benA*) or calmodulin (*caM*), has been recommended for the identification of individual species within sections [107]. Single-copy genes have also been used as molecular markers in the diagnosis of IA, which provide a greater specificity to the assay but a lower sensitivity. The single-copy genes that have been used include *aspHS*, *SCW4* and *anxC4*, which have shown promising results for detecting *A. fumigatus*, *A. flavus*, *A. niger*, *A. nidulans*, *A. terreus* and *A. versicolour* in BAL samples [93, 101, 106]. Because the *aspHS* gene encodes a haemolysin that is overexpressed *in vivo* during infection, its detection provides specificity to differentiate an active infection (germinated conidia) from a non-active one (non-germinated conidia) [101].

4.5 Resistance markers

PCR has also been used to detect resistance to antifungals, since an increasing number of environmental and clinical *A. fumigatus* isolates with a lower susceptibility to azoles have been observed. Infections caused by azole-resistant fungi are associated with a very high mortality rate. The available evidence suggests that resistance may be emerging as a result of the widespread use of these compounds in agricultural and clinical settings. The predominant resistance mechanism involves mutations in the *cyp51A* gene (L98H, Y121F, T289A and TR34), which encodes sterol demethylase, the target of azoles, and it has been shown that a variety of these mutations confer resistance to azoles. Molecular resistance tests have been used to assess fungal isolates as part of epidemiological surveillance studies of drug resistance and have been used to assess clinical isolates as a complement to phenotypic susceptibility testing. Real-time multiplex PCR has also been used to directly detect azole-resistant *Aspergillus* in BAL samples from patients at IA risk. However, more studies are needed to determine its potential utility in clinical care [107, 112].

5. Other diagnostic methods

Nowadays, there are some tests for the presumptive diagnosis of aspergillosis which, although are not specific for the identification of species, have been decisive in the case of IA. Among these are the Galactomannan test (GM), the Beta-D-glucan (BDG) assay and volatile organic compounds (VOCs) testing.

5.1 Galactomannans

The GM is a polysaccharide, the main component of the *Aspergillus* cell wall, which binds to and is released from the hyphae during growth [116, 117]. The galactomannan is not only found in the walls of *Aspergillus* but also in the cell walls of other fungi such as *Fusarium* spp., *Histoplasma capsulatum*, *Penicillium* spp., *Paecilomyces* spp. and to a lesser extent in other fungi [109, 118–121].

One of the most widely used tests for its determination is Platelia™ *Aspergillus* EIA (immunoenzymatic sandwich microplate assay), which uses monoclonal antibodies that are directed against the GM of *Aspergillus* [117, 122–124]. The real interest of this test in the diagnosis of aspergillosis lies in the fact that the galactomannan represents a good indirect indicator of the fungus [119] once it is released into the bloodstream; furthermore, it can be detected in body fluids such as serum, bronchoalveolar lavage (BAL), cerebrospinal fluid or pleural fluid [118].

Currently, it is considered as a serological method that facilitates the diagnosis of IA [125], even though it presents a highly variable sensitivity which ranges from 40 to 100% and depends on the population to be evaluated. Hachem et al. [118] showed in a study with patients suffering from haematologic malignancies associated to IA produced by *Aspergillus* non-*fumigatus* a sensitivity of 49%, whereas patients that presented IA by *A. fumigatus sensu stricto* showed a sensitivity of 13%, and the specificity was of 99% in both groups. Another study conducted by Maschmeyer et al. [126] reported that in patients undergoing chemotherapy for cancer or patients with hematopoietic stem cell transplantation, the sensitivity of the GM was of 67–100% and the specificity was of 86–99%. The meta-analysis conducted by Leeflang et al. [117] in patients with neutropaenia showed a sensitivity and a specificity of 78 and 81%, respectively. Pfeiffer et al. [116] found a sensitivity and a specificity of 95% in patients with hematopoietic stem cell transplant and solid organ transplants, respectively. It is important to highlight that the detection of GM in patients on antibacterial treatment results in the reduction of the test's specificity, while in patients on antifungal treatment, the sensitivity decreases in most cases. However, this does not occur when using caspofungin which increases the sensitivity [109, 127]. It is necessary to consider that when using bronchoalveolar lavage (BAL) to determine galactomannans, the test increases its sensitivity and turns out to be a more useful mortality prediction test [128].

5.2 BDG

The ß-glucans are glucose polymers (polysaccharides) of high-molecular weight that are found naturally in the cell wall of various organisms, such as bacteria, yeasts, fungi and plants. The BDG is produced by most fungi of medical importance such as *Candida* spp., *Aspergillus* spp. and *Pneumocystis jirovecii*. Among the fungi that release little BDG in serum, there can be found Mucorales fungi and *Cryptococcus* [129]. This antigen is important in fungal infections as it is released during the infection and has a higher sensitivity for the diagnosis of IA compared to GM and can be detected in the plasma of patients with mycosis. Therefore, this antigen can be used as a marker of fungal infection although it does not allow the identification of species [130] and is included as a diagnostic criterion in the European Organization for Research and Treatment of Cancer/Mycosis Study Group (EORTC/MSG) [90, 131].

For the detection of BDG, the Fungitell assay (Cape Cod Associates, Inc.) is used, which was approved by the Food and Drug Administration in 2003 for the presumptive diagnosis of IFI [132]. For this test, the results are variable in different studies, with sensitivity values ranging from 80 to 90% and specificity values from 36 to 92%, according to the cut-off value used [133, 134]. In studies with patients suffering from haematologic malignancies who develop IA, the sensibility of the test varies from 55 to 95%, and the specificity from 77 to 96% [133, 135]. It is important to mention that data from existing clinical studies for the Fungitell assay, which is the most widely used test today, suggest that the use of a detection limit of 80 pg/ml is associated with a greater precision than those with a result of 60–80 pg/ml, which are considered as indefinite. The above indicates that higher cut-off values dramatically decrease the sensitivity of the test, whilst increasing its specificity [135, 136].

The β-D-glucan assay is often useful in combination with culture, as it improves conditions for success [135]. Several factors that may increase the levels of BDG on IFI have been identified, such as thrombocyte infusions with leukocyte depletion filters, hemodialysis with cellulose membranes [137], the use of antibiotics such as amoxicillin-clavulanic acid or piperacillin-tazobactam [138], the use of surgical gauzes containing glucan, the administration of human blood products (immunoglobulins or albumin), severe mucositis and the presence of serious bacterial infections [129]. On the other hand, Pickering et al. [139] reported that high concentrations of bilirubin and triglycerides inhibit the levels of BDG and cause false-negative results, while hemolysis causes false-positive results.

5.3 Volatile organic compounds (VOCs)

The air exhaled by patients with invasive diseases contains a large number of VOCs, produced by different metabolic pathways, which can be used as biomarkers of pulmonary disease [109, 140]. Several techniques have been used to determine VOCs, like the gas chromatography and mass spectrometry which is impractical for use in the clinic [141–143]. One alternative seems to be the use of electronic noses and artificial olfactory systems that use a series of sensors that help discriminate each smell that represents a unique blend of VOCs, which function through pattern recognition algorithms called 'breathprints' [144, 145]. Several volatile organic compounds characteristic of *Aspergillus* spp. have been identified such as 3-octanone, isoamyl alcohol, ethanol, cyclohexanone, 2-methyl-2-propanol, 2-methylfuran, 2-ethyl-1-hexanol and 2-pentylfuran, among others [141–143, 146].

Among the advantages of these devices are their low cost, most of them are manual, easy to operate and provide results in few minutes. With regard to the benefits for the patient, they are non-invasive tests, safe, fast and easy to perform.

The sensitivity and specificity of this test is of 100 and 83.3%, respectively, which makes it one of the best options for the diagnosis of IA [140].

It is necessary to consider that in order to validate all these tests, more studies that provide additional advantages for their use must be conducted.

6. Discussion and conclusions

Currently, IA has become very important among fungal infections around the world since the number of cases has increased coupled with a high rate of morbidity and mortality, detected mainly in immunocompromised patients. The *A. fumigatus complex* has been reported as the most frequent cause of IA in immunocompromised hosts. During the last few years, new species have been described that belong to other complexes that also cause this disease, which represents a challenge to understand the epidemiology of this nosological entity. On the other hand, the appearance of new species causing IA generates problems to establish its diagnosis since conventional methods continue to be used, which fail to discriminate between closely related species, such as imaging, histopathology, microscopy and culture procedures. These last ones are still considered the gold standards even though they have a low sensitivity and are time-consuming [109]. Other methods that have been considered important are immunological methods to detect GM and BDG antigens in serum and other biological fluids. They also have limitations, such as cross reactions with other fungal species and interference with antibiotics such as b-lactams or with plasma infusion solutions [147]. In addition, these methods only identify the fungus at the complex level. It is important to identify the species, since recently, the presence of 'cryptic' *Aspergillus* species has been revealed in clinical samples of IA, which present differences in the susceptibility to antifungals, as is the case of voriconazole considered as the therapy of choice for invasive aspergillosis. The effectiveness of this drug is uncertain in the cryptic *Aspergillus* species, since it has been shown that resistance to multiple antifungal drugs is frequent, particularly in *A. lentulus*, *A. alliaceus*, *A. sydowii*, *A. calidoustus*, *A. keveii*, *A. insuetus* and *A. fumigatiaffinis* [19]. Currently, there is more awareness of the need to identify *Aspergillus* at the species level. This can be achieved through a polyphasic strategy [3]. In addition to the phenotypic characteristics, it includes the analysis of multilocus sequences, as well as PCR with specific probes for each species, such as multiplex qPCR, to identify clinically relevant *Aspergillus* species from the complex *A. fumigatus*, *A. terreus*, *A. flavus*, *A. niger and A. nidulans* [112]. However, molecular methods have not yet been recognized as diagnostic criteria for the identification of invasive fungal infection (IFI) by EORTC/MSG, due to the lack of standardization protocols, and the significant rates of false positives and false negatives. In addition, it should be considered that before its implementation in routine clinical practice, each diagnostic test must follow a long validation process, which involves various aspects such as limit of sensitivity, reproducibility and precision, so that the task is nothing simple; however, its use is paramount, so several recent studies have evaluated its application [108, 109, 148].

Therefore, it is clear that despite the efforts made so far to implement effective diagnostic methods, there is still no consensus about which is the ideal method. Therefore, as long as these methods are not standardized and their reliability is not guaranteed to improve the detection of *Aspergillus* spp. in an effective and timely manner, the diagnosis of IA will continue to represent a challenge.

Acknowledgements

This project was supported by PAPIIT-DGAPA (IT201318).

Author details

María del Rocío Reyes-Montes[1*], Esperanza Duarte-Escalante[1], María Guadalupe Frías-De-León[2], Erick Obed Martínez-Herrera[2] and Gustavo Acosta-Altamirano[2]

1 Departamento de Microbiología y Parasitología, Facultad de Medicina, Universidad Nacional Autónoma de México (UNAM), México Cd. MX, México

2 Hospital Regional de Alta Especialidad Ixtapaluca. Ixtapaluca, México

*Address all correspondence to: remoa@unam.mx

References

[1] Vanden Bossche H, Dromer F, Improvisi I, Lozano-Chiu M, Rex JH, Sanglard D. Antifungal drug resistance in pathogenic fungi. Medical Mycology. 1998;**36**(Suppl 1):119-128

[2] De Hoog GS, Guarro J, Gené J, Figueras MJ. Atlas of Clinical Fungi. 3rd ed. Utrecht: CBS-KNAW Fungal Biodiversity Center; 2011

[3] Samson RA, Varga J, Witiak SM, Geiser DM. The species concept in *Aspergillus*: Recommendations of an international panel. Studies in Mycology. 2007;**59**:71-73

[4] Loo VG, Bertrand C, Dixon C, Vityé D, DeSalis B, McLean AP, Brox A, Robson HG. Control of constructionassociated nosocomial Aspergillosis in an antiquated hematology unit. Infection Control and Hospital Epidemiology. 1996;**17**(6):360-364

[5] Wiederhold NP, Lewis RE, Kontoyiannis DP. Invasive aspergillosis in patients with hematologic malignancies. Pharmacotherapy. 2003;**23**(12):1592-1610

[6] Opal SM, Asp AA, Cannady PB Jr, Morse PL, Burton LJ, Hammer PG 2nd. Efficacy of infection control measures during a nosocomial outbreak of disseminated aspergillosis associated with hospital construction. The Journal of Infectious Diseases. 1986;**153**(3):634-637

[7] Denning DW. *Aspergillosis*. Schering-Plough Corporation. 2006. Website of *Aspergillus*: http://www.aspergillus.org.uk/indexhome.htm?library.php~main

[8] Pfaller MA, Diekema DJ. Epidemiology of invasive mycoses in North America. Critical Reviews in Microbiology. 2010;**36**(1):1-53

[9] Kousha M, Tadi R, Soubani AO. Pulmonary aspergillosis: A clinical review. European Respiratory Review. 2011;**20**(121):156-174

[10] Fortún J, Meije Y, Fresco G, Moreno S. Aspergillosis. Clinical forms and treatment. Enfermedades Infecciosas y Microbiología Clínica. 2012;**30**(4):201-208

[11] Steinbach WJ, Marr KA, Anaissie EJ, Azie N, Quan SP, Meier-Kriesche HU, Apewokin S, Horn DL. Clinical epidemiology of 960 patients with invasive aspergillosis from the PATH Alliance Registry. The Journal of Infection. 2012;**65**(5):453-464

[12] Tunnicliffe G, Schomberg L, Walsh S, Tinwell B, Harrison T, Chua F. Airway and parenchymal manifestations of pulmonary aspergillosis. Respiratory Medicine. 2013;**107**(8):1113-1123

[13] Geltner C, Lass-Flörl C. Invasive pulmonary aspergillosis in organ transplants—Focus on lung transplants. Respiratory Investigation. 2016;**54**(2):76-84

[14] Taccone FS, Van den Abeele AM, Bulpa P, Misset B, Meersseman W, Cardoso T, Paiva JA, Blasco-Navalpotro M, De Laere E, Dimopoulos G, Rello J, Vogelaers D, Blot SI, Study Investigators AICU. Epidemiology of invasive aspergillosis in critically ill patients: Clinical presentation, underlying conditions, and out comes. Critical Care. 2015;**19**:7

[15] von Eiff M, Roos N, Schulten R, Hesse M, Zühlsdorf M, van de Loo J. Pulmonary aspergillosis: Early diagnosis improves survival. Respiration. 1995;**62**(6):341-347

[16] Maertens JA, Nucci M, Donnelly JP. The role of antifungal treatment

in hematology. Haematologica. 2012;**97**(3):325-327

[17] Balajee SA, Kano R, Baddley JW, Moser SA, Marr KA, Alexander BD, Andes D, Kontoyiannis DP, Perrone G, Peterson S, Brandt ME, Pappas PG, Chiller T. Molecular identification of *Aspergillus* species collected for the transplant-associated infection surveillance network. Journal of Clinical Microbiology. 2009;**47**(10):3138-3141

[18] Alastruey-Izquierdo A, Mellado E, Peláez T, Pemán J, Zapico S, Alvarez M, Rodríguez-Tudela JL, Cuenca-Estrella M, FILPOP StudyGroup. Population-based survey of filamentous fungi and antifungal resistance in Spain (FILPOP Study). Anti-microbial Agents and Chemotherapy. 2013;**57**(7):3380-3387

[19] Alastruey-Izquierdo A, Alcazar-Fuoli L, Cuenca-Estrella M. Antifungal susceptibility profile of cryptic species of *Aspergillus*. Mycopathologia. 2014;**178**(5-6):427-433

[20] Escribano P, Peláez T, Muñoz P, Bouza E, Guinea J. Is azole resistance in *Aspergillus* fumigatus a problem in Spain? Antimicrobial Agents and Chemotherapy. 2013;**57**(6):2815-2820

[21] Chekiri-Talbi M, Denning DW. Estimation des infections fongiques en Algérie. Journal de Mycologie Médicale. 2017;**27**:139-145

[22] Gugnani HC, Denning WD, Rahim R, Sadat A, Belal M, Mahbub MS. Burden of serious fungal infections in Bangladesh. European Journal of Clinical Microbiology & Infectious Diseases. 2017;**36**:993-997

[23] Lagrou K, Maertens J, Van Even E, Denning DW. Burden of serious fungal infections in Belgium. Mycoses. 2015;**58**(Suppl. S5):1-5

[24] Giacomazzi J, Baethgen L, Carneiro LC, Millington MA, Denning DW, Colombo AL, Pasqualotto AC, in association with the LIFE Program. The burden of serious human fungal infections in Brazil. Mycoses. 2016;**59**:145-150

[25] Bamba S, Zida A, Sangaré I, Cissé M, Denning DW, Hennequin C. Burden of severe fungal infections in Burkina Faso. Journal of Fungi (Basel). 2018;**4**(1):35

[26] Mandengue CE, Denning DW. The burden of serious fungal infections in Cameroon. Journal of Fungi. 2018;**4**:44. DOI: 10.3390/jof4020044

[27] Dufresne SF, Cole DC, Denning DW, Sheppard DC. Serious fungal infections in Canada. European Journal of Clinical Microbiology & Infectious Diseases. 2017;**36**:987-992

[28] Alvarez-Duarte E, Denning DW. Serious fungal infections in Chile. European Journal of Clinical Microbiology & Infectious Diseases. 2017;**36**:983-986

[29] Alvarez-Moreno CA, Cortes JA, Denning DW. Burden of fungal infections in Colombia. Journal of Fungi (Basel). 2018;**4**(2):41

[30] Chrdle A, Mallátová N, Vašáková M, Haber J, Denning DW. Burden of serious fungal infections in the Czech Republic. Mycoses. 2015;**58**(Suppl. S5):6-14

[31] Zurita J, Denning DW, Paz-Y-Miño A, Solís MB, Arias LM. Serious fungal infections in Ecuador. European Journal of Clinical Microbiology & Infectious Diseases. 2017;**36**:975-981

[32] Zaki SM, Denning DW. Serious fungal infections in Egypt. European Journal of Clinical Microbiology & Infectious Diseases. 2017;**36**:971-974

[33] Gangneux JP, Bougnoux ME, Hennequin C, Godet C, Chandenier J, Denning DW, Dupont B, LIFE program, the Société française de mycologie médicale SFMM-study group. An estimation of burden of serious fungal infections in France. Journal de Mycologie Médicale. 2016;**26**:385-390

[34] Gamaletsou MN, Drogari-Apiranthitou M, Denning DW, Sipsas V. An estimate of the burden of serious fungal diseases in Greece. European Journal of Clinical Microbiology & Infectious Diseases. 2016;**35**:1115-1120

[35] Medina N, Samayoa B, Lau-Bonilla D, Denning DW, Herrera R, Mercado D, Guzmán B, Pérez JC, Arathoon E. Burden of serious fungal infections in Guatemala. European Journal of Clinical Microbiology & Infectious Diseases. 2017;**36**:965-969

[36] Sinkó J, Sulyok M, Denning DW. Burden of serious fungal diseases in Hungary. Mycoses. 2015;**58**(Suppl. S5):29-33

[37] Wadi J, Denning DW. Burden of serious fungal infections in Jordan. Journal of Fungi. 2018;**4**(1):15

[38] Kemaykin VM, Tabinbaev NB, Khudaibergenova MS, Olifirovich AA, Layzzat MA, Denning DW, Klimko N. An estimate of severe and chronic fungal diseases in the Republic of Kazakhstan. Journal of Fungi (Basel). 2018;**4**(1):34

[39] Huh K, Ha YE, Denning DW, Peck KR. Serious fungal infections in Korea. European Journal of Clinical Microbiology & Infectious Diseases. 2017;**36**:957-963

[40] Velayuthan RD, Samudi C, Singh HKL, Ng KP, Shankar EM, Denning DW. Estimation of the burden of serious human fungal infections in Malaysia. Journal of Fungi. 2018;**4**(1):38

[41] Corzo-León DE, Armstrong-James D, Denning DW. Burden of serious fungal infections in Mexico. Mycoses. 2015;**58**(Suppl. S5):34-44

[42] Khwakhali US, Denning DW. Burden of serious fungal infections in Nepal. Mycoses. 2015;**58**(Suppl. S5):45-50

[43] Nordoy I, Hesstvedt L, Andersen CT, Mylvaganam H, Kols NI, Falch BM, Tofteland S, Müller F, Denning DW. An estimate of the burden of fungal disease in Norway. Journal of Fungi (Basel). 2018;**4**(1):29

[44] Jabeen K, Farooqi J, Mirza S, Denning DW, Zafar A. Serious fungal infections in Pakistan. European Journal of Clinical Microbiology & Infectious Diseases. 2017;**36**:949-956

[45] Bustamante B, Denning DW, Campos PE. Serious fungal infections in Peru. European Journal of Clinical Microbiology & Infectious Diseases. 2017;**36**:943-948

[46] Sabino R, Verissímo C, Brandão J, Martins C, Alves D, Pais C, Denning DW. Serious fungal infections in Portugal. European Journal of Clinical Microbiology & Infectious Diseases. 2017;**36**:1345-1352

[47] Batac MCR, Denning DCW. Serious fungal infections in the Philippines. European Journal of Clinical Microbiology & Infectious Diseases. 2017;**36**:937-941

[48] Taj-Aldeen SJ, Chandra P, Denning DW. Burden of fungal infections in Qatar. Mycoses. 2015;**58**(Suppl. S5):51-57

[49] Mares M, Moroti-Constantinescu VR, Denning DW. The burden of fungal diseases in Romania. Journal of Fungi (Basel). 2018;**4**(1):31

[50] Klimko N, Kozlova Y, Khostelidi S, Shadrivova O, Borzova Y, Burygina E,

Vasilieva N, Denning DW. The burden of serious fungal diseases in Russia. Mycoses. 2015;**58**(Suppl. S5):58-62

[51] Rodriguez-Tudela JL, Alastruey-Izquierdo A, Gago S, Cuenca-Estrella M, León C, Miro JM, Nuñez Boluda A, Ruiz Camps I, Sole A, Denning DW, The University of Manchester in association with the LIFE program at http://www.LIFE-worldwide.org. Burden of serious fungal infections in Spain. Clinical Microbiology and Infection. 2015;**21**:183-189

[52] Chayakulkeeree M, Denning DW. Serious fungal infections in Thailand. European Journal of Clinical Microbiology & Infectious Diseases. 2017;**36**:931-935

[53] Faini D, Maokola W, Furrer H, Hatz C, Battegay M, Tanner M, Denning DW, Letang E. Burden of serious fungal infections in Tanzania. Mycoses. 2015;**58**(Suppl. S5):70-79

[54] Denning DW, Gugnani HC. Burden of serious fungal infections in Trinidad and Tobago. Mycoses. 2015;**58**(Suppl. S5):80-84

[55] Osmanov A, Denning DW. Burden of serious fungal infections in Ukraine. Mycoses. 2015;**58**(Suppl. S5):94-100

[56] Macedo-Viñas M, Denning DW. Estimating the burden of serious fungal infections in Uruguay. Journal of Fungi (Basel). 2018;**4**(1):37

[57] Pegorie M, Denning DW, Welfare W. Estimating the burden of invasive and serious fungal disease in the United Kingdom. The Journal of Infection. 2017;**74**(1):60-71

[58] Beardsley J, Denning DW, Chau NV, Yen NTB, Crump JA, Day JN. Estimating the burden of fungal disease in Vietnam. Mycoses. 2015;**58**(Suppl. S5):101-106

[59] Samson RA, Visagie CM, Houbraken J, Hong SB, Hubka V, Klaassen CHW, Perrone G, Seifert KA, Susca A, Tanney JB, Varga J, Kocsubé S, Szigeti G, Yaguchi T, Frisvad JC. Phylogeny, identification and nomenclature of the genus *Aspergillus*. Studies in Mycology. 2014;**78**:141-173

[60] Raper KB, Fennell DI. The Genus *Aspergillus*. Baltimore: Williams & Wilkins; 1965

[61] Gams W, Christensen M, Onions AH, Pitt JI, Samson RA. Infrageneric taxa of *Aspergillus*. In: Samson RA, Pitt JI, editors. Advances in *Penicillium* and *Aspergillus* Systematics. New York: Plenum Press; 1985. pp. 55-62

[62] Peterson SW. Phylogenetic analyses of *Aspergillus* species using DNA sequences from four loci. Mycologia. 2008;**100**(2):205-226

[63] Samson RA, Varga J. Molecular systematics of *Aspergillus* and its teleomorphs. In: Machida M, Gomi K, editors. *Aspergillus*: Molecular Biology and Genomics. 2010. pp. 19-40

[64] Varga J, Frisvad JC, Samson RA. Aspergillus sect. Aeni sect. nov., a new section of the genus for *A. karnatakaensis* sp. nov. and some allied fungi. IMA Fungus. 2010;**1**(12):197-205

[65] Peterson SW, Varga J, Frisvad JC, Samson RA. Phylogeny and subgeneric taxonomy of *Aspergillus*. In: Varga J, Samson RA, editors. *Aspergillus* in the Genomic Era. Wageningen: Wageningen Academic Publishers; 2008. pp. 33-56

[66] Houbraken J, Samson RA. Phylogeny of *Penicillium* and the segregation of Trichocomaceae into three families. Studies in Mycology. 2011;**70**(1):1-51

[67] Houbraken J, de Vries RP, Samson RA. Modern taxonomy of biotechnologically important *Aspergillus*

and *Penicillium* species. Advances in Applied Microbiology. 2014;**86**:199-249

[68] Hubka V, Nováková A, Kolařík M, Jurjević Ž, Peterson SW. Revision of *Aspergillus* section *Flavipedes*: Seven new species and proposal of section Jani sect. Nov. Mycologia. 2015;**107**(1):169-208

[69] Piontelli EL. Agentes comunes en las aspergilosis humanas: Conceptos primarios en la diferenciación de sus complejos de especies. Boletín Micológico. 2014;**29**:63-100

[70] Bongomin F, Gago S, Oladele RO, Denning DW. Global and multi-national prevalence of fungal diseases-estimate precision. Journal of Fungi. 2017;**3**(4):57

[71] Ullmann AJ, Aguado JM, Arikan-Akdagli S, Denning DW, Groll AH, Lagrou K, Lass-Flörl C, Lewis RE, Munoz P, Verweij PE, Warris A, Ader F, Akova M, Arendrup MC, Barnes RA, Beigelman-Aubry C, Blot S, Bouza E, Brüggemann RJM, Buchheidt D, Cadranel J, Castagnola E, Chakrabarti A, Cuenca-Estrella M, Dimopoulos G, Fortun J, Gangneux JP, Garbino J, Heinz WJ, Herbrecht R, Heussel CP, Kibbler CC, Klimko N, Kullberg BJ, Lange C, Lehrnbecher T, Löffler J, Lortholary O, Maertens J, Marchetti O, Meis JF, Pagano L, Ribaud P, Richardson M, Roilides E, Ruhnke M, Sanguinetti M, Sheppard DC, Sinkó J, Skiada A, Vehreschild MJGT, Viscoli C, Cornely OA. Diagnosis and management of *Aspergillus* diseases: Executive summary of the 2017 ESCMIDECMM-ERS guideline. Clinical Microbiology and Infection. 2017;**24**(S1):e1-e38

[72] Panackal AA, Imhof A, Hanley EW, Marr KA. *Aspergillus ustus* infections among transplant recipients. Emerging Infectious Diseases. 2006;**12**(3):403-408

[73] Verweij PE, Varga J, Houbraken J, Rijs AJMM, Lunel FMV, Blijlevens NMA, Shea YR, Holland SM, Warris A, Melchers WJG, Samson RA. *Emericella quadrilineata* as cause of invasive aspergillosis. Emerging Infectious Diseases. 2008;**14**(4):566-572

[74] Coelho D, Silva S, Vale-Silva L, Gomes H, Pinto E, Sarmento A, Pinheiro MD. *Aspergillus viridinutans*: An agent of adult chronic invasive aspergillosis. Medical Mycology. 2011;**49**(7):755-759

[75] Gyotoku H, Izumikawa K, Ikeda H, Takazono T, Morinaga Y, Nakamura S, Imamura Y, Nishino T, Miyazaki T, Kakeya H, Yamamoto Y, Yanagihara K, Yasuoka A, Yaguchi T, Ohno H, Miyzaki Y, Kamei K, Kanda T, Kohno S. A case of bronchial aspergillosis caused by *Aspergillus udagawae* and its mycological features. Medical Mycology. 2012;**50**(6):631-636

[76] Hubka V, Kubatova A, Mallatova N, Sedlacek P, Melichar J, Skorepova M, Mencl K, Lyskova P, Sramkova B, Chudickova M, Kolarik PHM. Rare and new etiological agents revealed among 178 clinical *Aspergillus* strains obtained from Czech patients and characterized by molecular sequencing. Medical Mycology. 2012;**50**(6):601-610

[77] Peláez T, Álvarez-Pérez S, Mellado E, Serrano D, Valerio M, Blanco JL, Garcia ME, Muñoz P, Cuenca-Estrella M, Bouza E. Invasive aspergillosis caused by cryptic *Asper-gillus* species: A report of two consecutive episodes in a patient with leukaemia. Journal of Medical Microbiology; **62**(Pt 3):474-478

[78] Yu J, Mu X, Li R. Invasive pulmonary aspergillosis due to *Emericella nidulans* var. *echinulata*, successfully cured by voriconazole and micafungin. Journal of Clinical Microbiology. 2013;**51**(4):1327-1329

[79] Bastos VR, Santos DW, Padovan AC, Melo AS, Mazzolin Mde A, Camargo LF, Colombo AL. Early invasive pulmonary sspergillosis in a kidney transplant recipient caused by *Aspergillus lentulus*:

First Brazilian report. Mycopathologia. 2015;**179**(3-4):299-305

[80] Yoshida H, Seki M, Umeyama T, Urai M, Kinjo Y, Nishi I, Toyokawa M, Kaneko Y, Ohno H, Miyazaki Y, Tomono K. Invasive pulmonary aspergillosis due to *Aspergillus lentulus*: Successful treatment of a liver transplant patient. Journal of Infection and Chemotherapy. 2015;**21**(6):479-481

[81] Zarrinfar H, Mirhendi H, Fata A, Khodadadi H, Kordbacheh P. Detection of *Aspergillus flavus* and *A. fumigatus* in bronchoalveolar lavage specimens of hematopoietic stem cell transplants and hematological malignancies patients by real-time polymerase chain reaction, nested PCR and mycological assays. Jundishapur Journal of Microbiology. 2015;**8**(1):e13744

[82] Masih A, Singh PK, Kathuria S, Agarwal K, Meis JF, Chowdharya A. Identification by molecular methods and matrix-assisted laser desorption ionization–time of flight mass spectrometry and antifungal susceptibility profiles of clinically significant rare *Aspergillus* species in a referral chest hospital in Delhi, India. Journal of Clinical Microbiology. 2016;**54**(9):2354-2364

[83] Taghizadeh-Armaki M, Hedayati MT, Moqarabzadeh V, Ansari S, Omran SM, Zarrin-far H, Saber S, Verweij PE, Denning DW, Seyedmousavi S. Effect of involved *Asper-gillus* species on galactomannan in bronchoalveolar lavage of patients with invasive aspergillosis. Journal of Medical Microbiology. 2017;**66**(7):898-904

[84] Zanganeh E, Hossein Zarrinfar H, Rezaeetalab F, Fata AM, Tohidi M, Najafzadeh MJ, Alizadeh M, Seyedmousavi S. Predominance of non-fumigatus *Aspergillus* species among patients suspected to pulmonary aspergillosis in a tropical and subtropical region of the Middle East. Microbial Pathogenesis. 2018;**116**:296-300

[85] Chrenkova V, Hubka V, Cetkovsky P, Kouba M, Weinbergerova B, Hornofova PLL, Hubacek P. Proven invasive pulmonary aspergillosis in stem cell transplant recipient due to *Aspergillus sublatus*, a cryptic species of *A. nidulans*. Mycopathologia. 2018;**183**(2):423-429

[86] Suleyman G, Alangaden GJ. Nosocomial fungal infections. Epidemiology, infection control, and prevention. Infectious Disease Clinics of North America. 2016;**30**(4):1023-1052

[87] Pana ZD, Roilides E, Warris A, Groll AH, Zaoutis T. Epidemiology of invasive fungal disease in children. Journal of the Pediatric Infectious Diseases Society. 2017;**6**(S1):S3-S11

[88] López-Medrano F, Fernández-Ruiz M, Silva JT, Carver PL, van Delden C, Merino E, Pérez-Saez MJ, Montero M, Coussement J, de Abreu Mazzolin M, Cervera C, Santos L, Sabé N, Scemla A, Cordero E, Cruzado-Vega L, Martín-Moreno PL, Len O, Rudas E, Ponce de León A, Arriola M, Lauzurica R, David MD, González-Rico C, Henríquez-Palop F, Fortún J, Nucci M, Manuel O, Paño-Pardo JR, Montejo M, Vena A, Sánchez-Sobrino B, Mazuecos A, Pascual J, Horcajada JP, Lecompte T, Moreno A, Carratalá J, Blanes M, Hernández D, Hernández-Méndez DA, Fariñas MC, Perelló-Carrascosa M, Muñoz P, Andrés A, Aguado JM, the Spanish Network for Research in Infectious Diseases (REIPI), the Group for the Study of Infection in Transplant Recipients (GESITRA) of the Spanish Society of Clinical Microbiology and Infectious Diseases (SEIMC), the Study Group for Infections in Compromised Hosts (ESGICH) of the European Society of Clinical Microbiology and Infectious Diseases (ESCMID), the Swiss Transplant Cohort Study (STCS). Multinational case-control study of risk factors for the development of

late invasive pulmonary aspergillosis following kidney transplantation. Clinical Microbiology and Infection. 2018;**24**(2):192-198

[89] Sönmez A, Eksi F, Mustafa P, Haydaroglu Sahin H. Investigating the presence of fungal agents in febrile neutropenic patients with hematological malignancies using different microbiological, serological, and molecular methods. Bosnian Journal of Basic Medical Sciences. 2015;**15**(3):40-47

[90] DePauw B, Walsh TJ, Donnelly JP, Stevens DA, Edwards JE, Calandra T, Pappas PG, Maertens J, Lortholary O, Kauffman CA, Denning DW, Patterson TF, Maschmeyer G, Bille J, Dismukes WE, Herbrecht R, Hope WW, Kibbler CC, Kullberg BJ, Marr KA, Muñoz P, Odds FC, Perfect JR, Restrepo A, Ruhnke M, Segal BH, Sobel JD, Sorrell TC, Viscoli C, Wingard JR, Zaoutis T, Bennett JE, European Organization for Research and Treatment of Cancer/Invasive Fungal Infections Cooperative Group; National Institute of Allergy and Infectious Diseases Mycoses Study Group (EORTC/MSG) Consensus Group. Revised definitions of invasive fungal disease from the European Organization for Research and Treatment of Cancer/Invasive Fungal Infections Cooperative Group and the National Institute of Allergy and Infectious Diseases Mycoses Study Group (EORTC/MSG) Consensus Group. Clinical Infectious Diseases. 2008;**46**(2):1813-1821

[91] Ambasta A, Carson J, Church DL. The use of biomarkers and molecular methods for the earlier diagnosis of invasive aspergillosis in immunocompromised patients. Medical Mycology. 2015;**53**(6):531-557

[92] Shah AA, Hazen KC. Diagnostic accuracy of histopathologic and cytopathologic examination of *Aspergillus* species. American Journal of Clinical Pathology. 2013;**139**(1):55-56

[93] Arancia S, Sandini S, De Carolis E, Vella A, Sanguinetti M, Norelli S, De Bernardis F. Use of SCW4 gene primers in PCR methods for the identification of six medically important *Aspergillus* species. The New Microbiologica. 2016;**39**(4):274-286

[94] Fernandez-Molina JV, Abad-Diaz-de-Cerio A, Sueiro-Olivares M, Pellon A, Ramirez-Garcia A, Garaizar J, Pemán J, Hernando FL, Rementeria A. Rapid and specific detection of section *Fumigati* and *Aspergillus fumigates* in human samples using a new multiplex real-time PCR. Diagnostic Microbiology and Infectious Disease. 2014;**80**:111-118

[95] Challier S, Boyer S, Abachin E, Berche P. Development of a serum-based Taqman real-time PCR Assay for diagnosis of invasive aspergillosis. Journal of Clinical Microbiology. 2004;**42**(2):844-846

[96] Hummel M, Spiess B, Kentouche K, Niggemann S, Böhm C, Reuter S, Kiehl M, Mörz H, Hehlmann R, Buchheidt D. Detection of *Aspergillus* DNA in cerebrospinal fluid from patients with cerebral aspergillosis by a nested PCR assay. Journal of Clinical Microbiology. 2006;**44**(11):3989-3993

[97] Faber J, Moritz N, Henninger N, Zepp F, Knuf M. Rapid detection of common pathogenic *Aspergillus* species by a novel real-time PCR approach. Mycoses. 2008;**52**(3):228-233

[98] Bernal-Martínez L, Gago S, Buitrago MJ, Gomez-Lopez A, Rodríguez-Tudela JL, Cuenca-Estrella M. Analysis of performance of a PCR-based assay to detect DNA of *Aspergillus fumigatus* in whole blood and serum: A comparative study with clinical samples. Journal of Clinical Microbiology. 2011;**49**(10):3596-3599

[99] Torelli R, Sanguinetti M, Moody A, Pagano L, Caira M, De Carolis E, Fuso L, De Pascale G, Bello G, Antonelli M, Fadda G, Posteraro B. Diagnosis of invasive aspergillosis by a commercial real-time PCR assay for *Aspergillus* DNA in bronchoalveolar lavage fluid samples from high-risk patients compared to a galactomannan enzyme immunoassay. Journal of Clinical Microbiology. 2011;**49**(12):4273-4278

[100] Walsh TJ, Wissel MC, Grantham KJ, Petraitiene R, Petraitis V, Kasai M, Francesconi A, Cotton MP, Hughes JE, Greene L, Bacher JD, Manna P, Salomoni M, Kleiboeker SB, Reddy SK. Molecular detection and species-specific identification of medically important *Aspergillus* species by real-time PCR in experimental invasive pulmonary aspergillosis. Journal of Clinical Microbiology. 2011;**49**(12):4150-4157

[101] Abad-Diaz-De-Cerio A, Fernandez-Molina JV, Ramirez-Garcia A, Sendino J, Hernando FL, Pemán J, Garaizar J, Rementeria A. The aspHS gene as a new target for detecting *Aspergillus fumigatus* during infections by quantitative real-time PCR. Medical Myco-logy. 2013;**51**(5):545-554

[102] Reinwald M, Spiess B, Heinz WJ, Heussel CP, Bertz H, Cornely OA, Hahn J, Lehrnbecher T, Kiehl M, Laws HJ, Wolf HH, Schwerdtfeger R, Schultheis B, Burchardt A, Klein M, Dürken M, Claus B, Schlegel F, Hummel M, Hofmann WK, Buchheidt D. *Aspergillus* PCR-based investigation of fresh tissue and effusion samples in patients with suspected invasive aspergillosis enhances diagnostic capabilities. Journal of Clinical Microbiology. 2013;**51**(12):4178-4185

[103] White PL, Barnes RA, Springer J, Klingspor L, Cuenca-Estrella M, Morton CO, Lagrou K, Bretagne S, Melchers WJG, Mengoli C, Donnelly JP, Heinz WJ, Loeffler J, the EAPCRI. Clinical performance of *Aspergillus* PCR for testing serum and plasma: A study by the European Aspergillus PCR Initiative. Journal of Clinical Microbiology. 2015;**53**(9):2832-2837

[104] White PL, Posso RB, Barnes RA. Analytical and clinical evaluation of the PathoNostics AsperGenius assay for detection of invasive aspergillosis and resistance to azole antifungal drugs directly from plasma samples. Journal of Clinical Microbiology. 2017;**55**(8):2356-2366

[105] Grancini A, Orlandi A, Lunghi G, Consonni D, Pozzi C, Rossetti V, Palleschi A, Fracchiolla N, Melada E, Savioli M, Arghittu M, Maiavacca R, Prigitano A. Evaluation of real time PCR *Aspergillus* spp. in bronchoalveolar lavage samples. The New Microbiologica. 2018;**41**(1): 67-70

[106] Tang Q, Tian S, Yu N, Zhang X, Jia X, Zhai H, Sun Q, Han L. Development and evaluation of a loop-mediated isothermal amplification method for rapid detection of *Aspergillus fumigatus*. Journal of Clinical Microbiology. 2016;**54**(4):950-955

[107] Powers-Fletcher MV, Hanson KE. Molecular diagnostic testing for *Aspergillus*. Journal of Clinical Microbiology. 2016;**54**(11):2655-2660

[108] White PL, Bretagne S, Klingspor L, Melchers WJ, McCulloch E, Schulz B, Finnstrom N, Mengoli C, Barnes RA, Donnelly JP, Loeffler J, European Aspergillus PCR Initiative. *Aspergillus* PCR: One step closer to standardization. Journal of Clinical Microbiology. 2010;**48**(4):1231-1240

[109] Arvanitis M, Anagnostou T, Burgwyn Fuchs B, Caliendo AM, Mylonakis E. Molecular and nonmolecular diagnostic methods for invasive fungal infections.

Clinical Microbiology Reviews. 2014;**27**(3):490-526

[110] Suarez F, Lortholary O, Buland S, Rubio MT, Ghez D, Mahé V, Quesne G, Poirée S, Buzyn A, Varet B, Berche P, Bougnoux ME. Detection of circulating *Aspergillus fumigatus* DNA by real-time PCR assay of large serum volumes improves early diagnosis of invasive aspergillosis in high-risk adult patients under hematologic surveillance. Journal of Clinical Microbiology. 2008;**46**(11):3772-3777

[111] Springer J, White PL, Hamilton S, Michel D, Barnes RA, Einsele H, Loffler J. Comparison of performance characteristics of *Aspergillus* PCR in testing a range of blood-based samples in accordance with international methodological recommendations. Journal of Clinical Microbiology. 2016;**54**(3):705-711

[112] Chong GLM, van de Sande WWJ, Dingemans GJH, Gaajetaan GR, Vonk AG, Hayette MP, van Tegelen DWE, Simons GFM, Rijnders BJA. Validation of a new *Aspergillus* real-time PCR assay for direct detection of *Aspergillus* and azole resistance of *Aspergillus fumigatus* on bronchoalveolar lavage fluid. Journal of Clinical Microbiology. 2015;**53**(3):868-874

[113] Mikulska M, Furfaro E, Viscoli C. Non-cultural methods for the diagnosis of invasive fungal disease. Expert Review of Anti-Infective Therapy. 2015;**13**(1):103-117

[114] Alanio A, Menotti J, Gits-Muselli M, Hamane S, Denis B, Rafoux E, Peffault de la Tour R, Touratier S, Bergeron A, Guigue N, Bretagne S. Circulating *Aspergillus fumigatus* DNA is quantitatively correlated to galactomannan in serum. Frontiers in Microbiology. 2017;**8**:2040

[115] Schoch CL, Seifert KA, Huhndorf S, Robert V, Spouge JL, Levesque CA, Chen W. Nuclear ribosomal internal transcribed spacer (ITS) region as a universal DNA barcode marker for fungi. Proceedings of the National Academy of Sciences of the United States of America. 2012;**109**(16):6241-6246

[116] Pfeiffer CD, Fine JP, Safdar N. Diagnosis of invasive aspergillosis using a galactomannan assay: A meta-analysis. Clinical Infectious Diseases. 2006;**42**(10):1417-1427

[117] Leeflang MM, Debets-Ossenkopp YJ, Visser CE, Scholten RJ, Hooft L, Bijlmer HA, Reitsma JB, Bossuyt PM, Vandenbroucke-Grauls C. Galactomannan detection for invasive aspergillosis in immunocompromised patients. Cochrane Database of Systematic Reviews. 2008;**12**:CD007394

[118] Hachem RY, Kontoyiannis DP, Chemaly RF, Jiang Y, Reitzel R, Raad I. Utility of galactomannan enzyme immunoassay and (1,3) beta- D-glucan in diagnosis of invasive fungal infections: Low sensitivity for *Aspergillus fumigatus* infection in hematologic malignancy patients. Journal of Clinical Microbiology. 2009;**47**(1):129-133

[119] Quindós G. New microbiological techniques for the diagnosis of invasive mycoses caused by filamentous fungi. European Journal of Clinical Microbiology & Infectious Diseases. 2006;**12**(Suppl 7):40-52

[120] Tortorano AM, Esposto MC, Prigitano A, Grancini A, Ossi C, Cavanna C, Cascio G. Cross-reactivity of *Fusarium* spp. in the *Aspergillus* galactomannan enzyme-linked immunosorbent assay. Journal of Clinical Microbiology. 2012;**50**(3):1051-1053

[121] Wheat LJ, Hackett E, Durkin M, Connolly P, Petraitiene R, Walsh TJ, Knox K, Hage C. Histoplasmosis-associated cross-reactivity in the BioRad Platelia *Aspergillus* enzyme

immunoassay. Clinical and Vaccine Immunology. 2007;**14**(5):638-640

[122] Latgé JP, Kobayashi H, Debeaupuis JP, Diaquin M, Sarfati J, Wieruszeski JM. Chemical and immunological characterization of the extracellular galactomannan of *Aspergillus fumigatus*. Infection and Immunity. 1994;**62**(12):5424-5433

[123] Stynen D, Sarfati J, Goris A, Prévost MC, Lesourd M, Kamphuis H, Darras V, Latgé JP. Rat monoclonal antibodies against *Aspergillus* galactomannan. Infection and Im-munity. 1992;**60**(6):2237-2245

[124] Stynen D, Goris A, Sarfati J, Latge JP. A new sensitive sandwich enzymelinked immunosorbent assay to detect galactofuran in patients with invasive aspergillosis. Journal of Clinical Microbiology. 1995;**33**(2):497-500

[125] Mennink-Kersten MA, Donnelly JP, Verweij PE. Detection of circulating galactomannan for the diagnosis and management of invasive aspergillosis. The Lancet Infectious Diseases. 2004;**4**(6):349-357

[126] Maschmeyer G, Calandra T, Singh T, Wiley J, Perfect J. Invasive mould infections: A multi-disciplinary update. Medical Mycology. 2009;**47**(6):571-583

[127] Klont RR, Mennink-Kersten MA, Ruegebrink D, Rijs AJ, Blijlevens NM, Donnelly JP, Verweij PE. Paradoxical increase in circulating *Aspergillus* antigen during treatment with caspofungin in a patient with pulmonary aspergillosis. Clinical Infectious Diseases. 2006;**43**(3): e23-e25

[128] Zou M, Tang L, Zhao S, Zhao Z, Chen L, Chen P, Huang Z, Li J, Chen L, Fan X. Systematic review and meta-analysis of detecting galactomannan in bronchoalveolar lavage fluid for diagnosing invasive aspergillosis. PLoS One. 2012;7(8):e43347

[129] Sulahian A, Porcher R, Bergeron A, Touratier S, Raffoux E, Menotti J, Derouin F, Ribaud P. Use and limits of (1-3)-β-D-glucan assay (Fungitell), compared to galactomannan determination (Platelia *Aspergillus*), for diagnosis of invasive sspergillosis. Journal of Clinical Microbiology. 2014;**52**(7):2328-2333

[130] Pizarro S, Ronco AM, Gotteland M. β-glucanos: ¿qué tipos existen y cuáles son sus beneficios en la salud? Revista Chilena de Nutricion. 2014;**41**(3):439-446

[131] De Pauw B, Walsh TJ, Donnally JP, Stenvens DA, Edwards JE, Calandra T. Revised definitions of invasive fungal disease from the European Organization for Research and Treatment of Cancer/ Invasive Fungal Infections Cooperative Group and the National Institute of Allergy and Infectious Diseases Mycoses Study Group (EORTC/MSG) Consensus Group. Clinical Infectious Diseases. 2008;**46**(12):1813-1821

[132] Theel ES, Doern CD. β-D-Glucan testing is important for diagnosis of invasive fungal infections. Journal of Clinical Microbiology. 2013;**51**(11):3478-3483

[133] Ostrosky-Zeichner L, Alexander BD, Kett DH, Vasquez J, Pappas PG, Saek F, Ketchum PA, Wingard J, Schiff R, Tamura H, Finkelman MA, Rex JH. Multicenter clinical evaluation of the (1→3)-ß-D-Glucan assay as an aid to diagnosis of fungal infections in humans. Clinical Infectious Diseases. 2005;**41**(5):654-659

[134] De Vlieger G, Lagrou K, Maertens J, Verbeken E, Meersseman W, Van Wijngaerden E. (1→3)-ß-D-Glucan detection as a diagnostic test for invasive aspergillosis in immunocompromised critically ill patients with symptoms of respiratory infection: An autopsy-based study. Journal of Clinical Microbiology. 2011;**49**(11):3783-3787

[135] Lamoth F, Cruciani M, Mengoli C, Castagnola E, Lortholary O, Richardson M, Marchetti O. Third European Conference on Infections in Leukemia (ECIL-3). ß-Glucan antigenemia assay for the diagnosis of invasive fungal infections in patients with hematological malignancies: A systematic review and meta-analysis of cohort studies from the Third European Conference on Infections in Leukemia (ECIL-3). Clinical Infectious Diseases. 2012;**54**(5):633-643

[136] Obayashi T, Negishi K, Suzuki T, Funata N. Reappraisal of the serum (13)-beta-D-glucan assay for the diagnosis of invasive fungal infections—A study based on autopsy cases from 6 years. Clinical Infectious Diseases. 2008;**46**(12):1864-1870

[137] Kanda H, Kubo K, Hamasak K, Kanda Y, Nakao A, Kitamura T, Fujita T, Yamamoto K, Mimura T. Influence of various hemodialysis membranes on the plasma (1-3)-β-d-glucan level. Kidney International. 2001;**60**(1):319-323

[138] Marty FM, Lowry CM, Lempitski SJ, Kubiak DW, Finkelman MA, Baden LR. Reacti-vity of (1-3)-β-D-glucan assay with commonly used intravenous antimicrobials. Anti-microbial Agents and Chemotherapy. 2006;**50**(10):3450-3453

[139] Pickering JW, Sant HW, Bowles CAP, Roberts WL, Woods GL. Evaluation of a (1→3)-ß-D-Glucan assay for diagnosis of invasive fungal infections. Journal of Clinical Microbiology. 2005;**43**(12):5957-5962

[140] de Heer K, van der Schee MP, Zwinderman K, van den Berk IA, Visser CE, van Oers R, Sterk PJ. Electronic nose technology for detection of invasive pulmonary aspergillosis in prolonged chemotherapy-induced neutropenia: A proof-of-principle study. Journal of Clinical Microbiology. 2013;**51**(5):1490-1495

[141] Chambers ST, Syhre M, Murdoch DR, McCartin F, Epton MJ. Detection of 2-pentylfuran in the breath of patients with *Aspergillus fumigatus*. Medical Mycology. 2009;**47**(5):468-476

[142] Gerritsen MG, Brinkman P, Escobar N, Bos LD, de Heer K, Meijer MM. Profiling of volatile organic compounds produced by clinical *Aspergillus* isolates using gas chromatography–mass spectrometry. Medical Mycology. 2018;**56**(2):253-256

[143] Syhre M, Scotter JM, Chambers ST. Investigation into the production of 2-pentylfuran by *Aspergillus fumigatus* and other respiratory pathogens in vitro and human breath samples. Medical Mycology. 2008;**46**(3):209-215

[144] Lewis NS. Comparisons between mammalian and artificial olfaction based on arrays of carbon black-polymer composite vapor detectors. Accounts of Chemical Research. 2004;**37**(9):663-672

[145] Wilson AD, Baietto M. Advances in electronic-nose technologies developed for biomedical applications. Sensors (Basel). 2011;**11**(1):1105-1176

[146] Perl T, Jünger M, Vautz W, Nolte J, Kuhns M, Zepelin MB, Quintel M. Detection of characteristic metabolites of *Aspergillus fumigatus* and *Candida* species using ion mobility spectrometry–metabolic profiling by volatile organic compounds. Mycoses. 2011;**54**(6):e828-e837

[147] Barton RC. Laboratory diagnosis of invasive aspergillosis: From diagnosis to prediction of outcome. Scientifica. 2013;**2013**:29

[148] Ibáñez-Martínez E, Ruiz-Gaitán A, Pemán-García J. Update on the diagnosis of invasive fungal infection. Revista Española de Quimioterapia. 2017;**30**(Suppl 1):16-21

Chapter 8

Metabolomics: Basic Principles and Strategies

Sinem Nalbantoglu

Abstract

Metabolomics is the study of metabolome within cells, biofluids, tissues, or organisms to comprehensively identify and quantify all endogenous and exogenous low-molecular-weight (<1 kDa) small molecules/metabolites in a biological system in a high-throughput manner. Metabolomics has several applications in health and disease including precision/personalized medicine, single cell, epidemiologic population studies, metabolic phenotyping, and metabolome-wide association studies (MWAS), precision metabolomics, and in combination with other omics disciplines as integrative omics, biotechnology, and bioengineering. Mass spectrometry (MS)-based metabolomics/lipidomics provides a useful approach for both identification of disease-related metabolites in biofluids or tissue and also encompasses classification and/or characterization of disease or treatment-associated molecular patterns generated from metabolites. Here, in this review, we provide a brief overview of the current status of promising MS-based metabolomics strategies and their emerging roles, as well as possible challenges.

Keywords: metabolomics, untargeted metabolomics, targeted metabolomics, omics, mass spectrometry

1. Introduction

Metabolomics is an evolving field to comprehensively identify and quantify all endogenous and exogenous low-molecular-weight (<1 kDa) small molecules/metabolites in a biological system in a high-throughput manner. The composition of these endogenous compounds is affected by the upstream influence of the proteome and genome as well as environmental factors, lifestyle factors, medication, and underlying disease. Metabolomics is reported as the reflection of the phenotype. The metabolome is downstream of the transcriptome and proteome and is considered to be complementary to genomics, transcriptomics, and proteomics. It has been reported that due to the close relation of metabolome to the genotype, physiology, and environment of an organism, genotype-phenotype as well as genotype-envirotype relationships could be successfully documented by metabolomics [1–3].

Metabolomics is the study of metabolome within cells, biofluids, tissues, or organisms and applied in molecular and personalized medicine involved in clinical chemistry, transplant monitoring, newborn screening, pharmacology, and toxicology. Metabolome can be defined as the small molecules and their interactions within a biological system which has been estimated as 3000–20,000 global metabolite profiles under a given genetic, nutritional, environmental conditions. Since the metabolome is the final downstream product, changes and interactions between gene expression, protein expression, and the environment are directly reflected in metabolome making it more physically and chemically complex than the other "omes." The metabolome is closest to the phenotype among other omics approaches, and metabolomics best modulates and represents the molecular phenotype of health and disease [4]. It has been demonstrated that relations of genotype-genomics and phenotype-metabolomics refer to specific gene variations and resultant metabolite changes which ultimately give information about genetic epigenetic phenotypic changes [1, 5–8]. In this regard, metabolomics is a brilliant source for biomarker discovery with advantages over other omics approaches.

An overview of first metabolomics experiments has demonstrated metabolite quantitation in biofluids which was first utilized in 1971 [9, 10]. After that, in the same year, the first use of the definition of "metabolic profiling" was observed [9, 11], while "metabolome" was first used in 1998 [9, 12]. In 1999, the term "metabonomics" was described as the quantitative measurement of the dynamic multiparametric metabolic response of living systems to pathophysiological stimuli or genetic modification [13].

Due to the complexity of the metabolome, a wide variety of chemically diverse compounds such as lipids, organic acids, carbohydrates, amino acids, nucleotides, and steroids, among others, have not yet been validated and reported totally; thus, we do not know the complete number of metabolites present in the human [14]. Human Metabolome Database (HMDB; http://www.hmdb.ca, version 4.0, 2018) reported 114,098 (April 2019) metabolite entries including both water-soluble and lipid-soluble metabolites as well as metabolites that would be regarded as either abundant (>1 uM) or relatively rare (<1 nM). Those small molecules, which have been identified and experimentally confirmed in various human tissues and biofluids, have been suggested as only 20% of the total metabolome [15]. Additionally, 5702 protein sequences are linked to these metabolite entries. The database also contained listing normal and abnormal concentrations of different metabolites for 23 different biospecimens.

Metabolomics strategies cover two primary analysis platforms including "untargeted-discovery-global" and "targeted-validation-tandem" based on the objective of the study (**Figure 1**). In order to systematically identify and quantify metabolites from a biological sample and achieve comprehensive characterization of biomarker targets, the analysis considers both endometabolome and exometabolome. Untargeted discovery metabolomics has a hypothesis-generating manner and allows for full scanning of the metabolome, pattern identification, and "metabolic fingerprinting" for the global classification of phenotypes with interacting pathway interactions. Targeted metabolomics is hypothesis testing and generally performed for validation of an untargeted analysis. In the targeted approaches (tandem-MS/MS), using a known standard, a quantitative analysis is performed on specific small molecules/metabolites or perturbations along a metabolic pathway [9] also known as "biased or directed metabolomics" or "metabolic profiling." Hypothesis-generating metabolomics covers different strategies as (i) nontargeted profiling, (ii) fingerprinting, and (iii) footprinting [16], while hypothesis-testing strategies are target analysis and diagnostic analysis. Nontargeted global metabolomics profiling refers to comprehensive metabolite/small molecule analysis. This analysis performs semiquantitative analysis with putative identifications of the detected features. Metabolomics fingerprint examines global snapshot of the intracellular metabolome enabling classification and screening, while metabolomics footprint analysis explores global snapshot of the extracellular fluid metabolome (secretions from cells or changes in metabolites consumed from the exometabolome). Quantification and identification of the

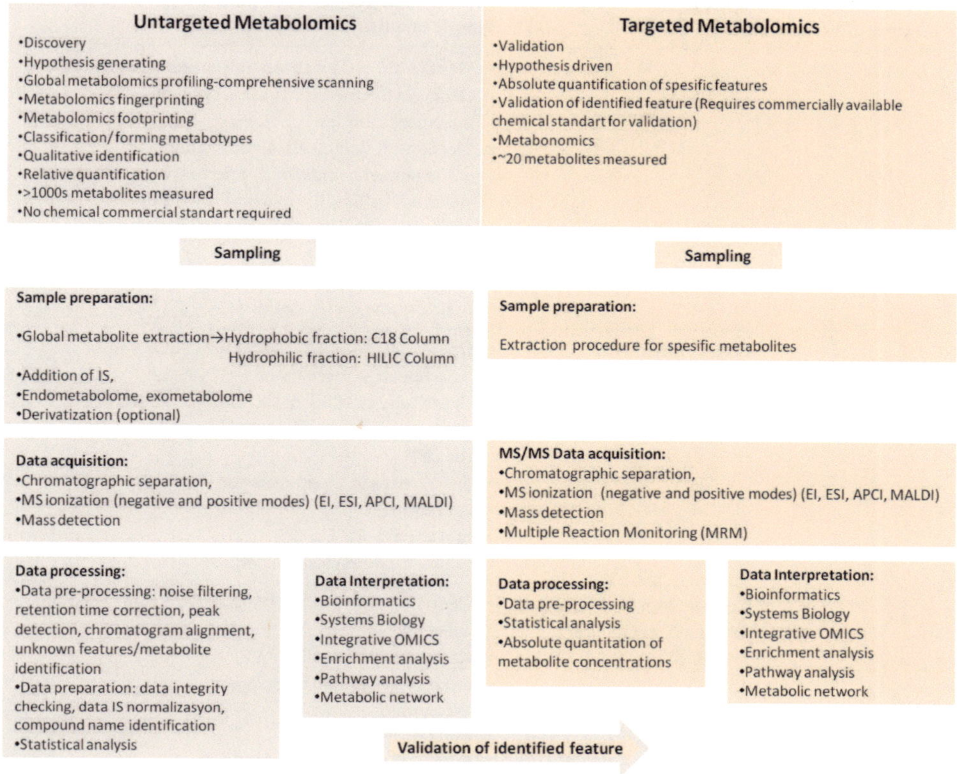

Figure 1.
Metabolomics workflow with main bulleted points.

features do not comply with fingerprint and footprint metabolomics. Metabolomics strategies for validation purposes refer to quantitative tandem/targeted analysis and diagnostic analysis of a known clinical associated compound/biomarker [16]. Untargeted and targeted approaches should be performed consecutively in order to achieve an accurate identification and absolute quantitation of the metabolites [9]. Here, in this review, we provide a brief overview of the current status of promising MS-based metabolomics strategies and their emerging roles, as well as possible challenges.

2. Basic workflow of MS-based metabolomics

MS is used to identify and quantify metabolites even at very low concentrations (femtomolar to attomolar) with high resolution, sensitivity, and dynamic range [17]. MS-based analyses basically include sample preparation, extraction, capillary electrophoresis (CE), and/or chromatographic separation, introduction of sample for ionization process (charged molecules), and detection of possible metabolites on the basis of their mass-to-charge ratio (m/z). In this review, the main aspects of MS-based untargeted metabolomics strategies are briefly outlined below:

1. **Sample acquisition:** Metabolome analysis can be performed in various biological samples including tissue, biofluids (blood, urine, feces, seminal fluid, saliva, bile, cerebrospinal fluid), and cell culture [18] (**Table 1**). Careful sampling, sample preparation and management, and sample biobanking/biorepositories together with sample labeling are critical and essential for an

Types of biospecimen		Sample acquisition and preparation
Tissue		• Mechanical and nonmechanical homogenization can be performed for tissues, cells, and other biological samples • Mechanical homogenization include homogenizer, ultrasound, microwave, manual grinding, ball mill, and grinding in a liquid nitrogen-cooled mortar and pestle • Aliquoting • Storage at −80°C • Freeze-thaw cycles avoided
Suspension-cultured mammalian cells	Cells (intracellular-fingerprint-metabolite profiling)	• Quenching with ice-cold 50 mM ammonium bicarbonate + methanol • Centrifuge at 1000×*g* for 1 min at −20°C • Removal of the media/quenching solution from the cells • Snap freeze in liquid nitrogen • Aliquoting • Storage at −80°C • Freeze-thaw cycles avoided
	Cell medium (extracellular-footprint-metabolite profiling)	• Centrifuge the media at 500×*g* for 5 min (remove cells) • Remove supernatant • Snap freeze in liquid nitrogen • Aliquoting • Storage at −80°C • Freeze-thaw cycles avoided
Biofluids	Serum	• Blood samples should be collected into serum separator tubes and incubate 30 min and no longer than 60 min for clotting procedure on ice instead of room temperature to minimize residual metabolic activity • At the end of the clotting time, the blood sample is centrifuged or 20 min at 1100–1300×*g* at ambient temperature • The serum visible in the upper layer of the tube as supernatant is collected and stored at −80°C, and freeze-thaw cycles are avoided
	Plasma	• Blood samples collected into heparin, citrate, or EDTA-containing tubes under fasting conditions • Centrifuge at 13,000×*g*, 15 min, 4°C • Supernatant isolation and removal • Aliquoting • Storage at −80°C • Freeze-thaw cycles avoided • Various coagulants such as heparin, citrate, or EDTA represent different retention time peaks
	Cerebrospinal fluid (CSF)	• CSF sampling via lumbar puncture • Aliquoting • Storage at −80°C • Freeze-thaw cycles avoided • To obtain highly resolved chromatograms, careful handling is essential in case of blood-contaminated CSF sample
	Urine	• Sodium azide addition to the sample as bacteriostatic agent during sample storage • 0.2 μm filtration for each sample in order to avoid any particulates

Types of biospecimen	Sample acquisition and preparation
	• Centrifuge at 1000–3000 rcf for 5 min • Removing supernatant • Aliquoting • Storage at −80°C • Freeze-thaw cycles avoided
Saliva	• 3 mL saliva samples collected into sample tubes under fasting conditions • Aliquoting • Storage at −80°C
Sweat	• Sweat secretion is stimulated by exercise, heat, or chemicals such as pilocarpine • The skin area is cleaned with ethanol and then with distilled water • The sweat is collected by using a micropipette (from 20 to 200 μL at least 100 μL) • The minimum sweat rate demanded to obtain a valid sweat sampling is 1 g/m^2 per min • The use of deodorants, perfumes, and cosmetics was excluded at least 1 day prior to sweat collection • Aliquoting • Storage at −80°C • Freeze-thaw cycles avoided
Feces	• A single spot 1–2 g of feces • Storage at −80°C
Breath	• Exhaled breath condensate (EBC) upon 8 h of fasting including smoking is performed • Collecting time for EBC is 20 min, producing approximately 1 mL condensate sample • Commercial collection samplers, bags, and devices are used • Compatible with GC, GC-MS • Aliquoting • Storage at −80°C • Freeze-thaw cycles avoided
Seminal plasma	• Semen is centrifuged at 700 × *g*, 4°C, 10 min • Supernatant (seminal plasma) is separated • Supernatant seminal plasma is recentrifuged at 10,000 × *g*, 60 min, 4°C • Aliquoting • Storage at −80°C • Freeze-thaw cycles avoided
Bile	• 2 mL gallbladder bile is collected intraoperatively • Aliquoting • Storage at −80°C • Freeze-thaw cycles avoided

Table 1.
Systematic collection and storage of human samples for metabolomics/lipidomics strategies.

optimum, reproducible, and high-throughput analysis with high recovery, extraction, and enriched metabolite coverage.

2. **Sample preparation/extraction:** During sample preparation different extraction solvents/methods are used for high recovery of both polar and nonpolar compounds based on nontargeted and targeted approaches on

different biological samples involving tissue, blood, plasma, serum, cells, urine, etc. Basically applied approaches include optimized methanol-water-chloroform combinations to extract both hydrophilic and hydrophobic compounds. For high recovery of both hydrophilic and hydrophobic compounds, separate extraction applications give better results. During sample preparation after the centrifugation process, a biphasic mixture of the upper (aqueous) and lower (organic) layers is extracted separately. In the two sequential or two-phase extraction applications optimized for both polar and nonpolar metabolites such as lipids, an aqueous extraction using polar organic solvents (e.g., methanol or acetonitrile) mixed with water followed by organic extraction (lipid extraction) with dichloromethane or chloroform-methanol is carried out [16, 19]. The first one of the two-phase extractions involves aqueous solvent (e.g., methanol-water) followed by extraction with a nonpolar solvent (e.g., chloroform) of the centrifuged pellet.

3. **Separation:** Chromatographic separation techniques including liquid chromatography (LC) and gas chromatography (GC) are used coupling to MS systems (GC-MS, HPLC-MS, UPLC-MS), while direct injection techniques include direct infusion MS [20] and direct analysis in real-time MS (DART-MS) [21, 22]. In addition, capillary electrophoresis (CE) coupled to MS systems (CE-MS) is an important technique for separation and profiling of polar metabolites in biological samples. Reversed-phase LC using C18 columns is used for separation of nonpolar compounds, while hydrophilic interaction chromatography (HILIC) is used for separation of polar compounds [23].

Gas chromatography/mass spectrometry (GC/MS) is one of the widely used untargeted and targeted metabolomics platforms which offer high chromatographic resolution. In addition to other compounds, volatile organic compounds (VOCs) such as fatty acids and organic acids which are important biomarker candidates in biological samples can be successfully achieved by GC-MS. Analysis by GC-MS requires derivatization with reactions of alkylation, acylation, and silylation in order to increase detection or retention of the compound [24].

Volatile organic compounds are important components of the metabolome and include metabolites such as alcohols, alkanes, aldehydes, furans, ketones, pyrroles, and terpenes. Volatilomics is a new field with adductomics into the metabolomics. For the extraction of VOCs, solvent-free sample preparation/extraction method “solid-phase microextraction (SPME)” is used, which enables extraction of organic compounds from gaseous, aqueous, and solid materials [24].

LC/MS and GC-MS have different sensitivities for detecting metabolites with high recovery/coverage. Compared to GC/MS, a wide range of molecular features can be analyzed via LC-MS. GC-MS is capable of analyzing less polar biomolecules involving alkylsilyl derivatives, eicosanoids, essential oils, esters, perfumes, terpenes, waxes, volatiles, carotenoids, flavonoids, and lipids. LC-MS is capable of analyzing more polar biomolecules involving organic acids, organic amines, nucleosides, ionic species, nucleotides, and polyamines and does not require derivatization. Both LC-MS and GC-MS are able to analyze alcohols, alkaloids, amino acids, catecholamines, fatty acids, phenolics, polar organics, prostaglandins, and steroids [25].

4. **Ionization:** After chromatographic separation, samples are pumped through MS capillary to obtain positive or negative electrically charged ions in gas phase. Introduction of heat and dry nitrogen in the MS cause the droplets to

evaporate. Then the resultant evaporated droplets transfer the charge to the analytes and ionize them both in the positive and negative mode via charge transfer [26]. Polarity of the ionization/ion sources has great importance for avoiding metabolites' losses. Based on the polarity of the molecule, applied ionization sources include electron ionization (EI), chemical ionization (CI), electrospray ionization source (ESI), atmospheric pressure chemical ionization (APCI), atmospheric pressure photo-ionization (APPI), and matrix assisted laser desorption ionization (MALDI) [27–31].

5. **Detection:** High-resolution mass spectrum composed of mass-to-charge (m/z) ratios of fragment ions created by ionized biomolecules is detected by MS at sub-femtomole levels. Mass analyzers include time of flight (TOF), quadrupole time of flight (QTOF), quadrupole, ion trap, and orbitrap. For targeted metabolomics analysis, tandem or MS/MS is performed for validation of potentially discovered metabolites during untargeted analysis. For MS/MS, ion trap or triple quadrupole (QQQ) with multiple reaction monitoring (MRM) is generally used with high sensitivity, mass resolution, and accuracy (<1 ppm mass error) [9].

6. **Data analysis and metabolite identification:** The large amounts of complex raw data involving specific metabolic signals are extracted from MS and analyzed in specialized software to properly interpret the data and identify the metabolite of interest. Commercially available and free software bioinformatic analysis tools automatically perform processing of peak selection, assessment, and relative quantitation. Raw data signal spectrum preprocessing includes background spectral filtering (noise elimination), retention time correction, appropriate peak assignment for the same compound (identification of matching m/z and assigning adducts appropriately), peak detection, peak alignment (matching peaks across multiple samples) and peak normalization (adjusting peak intensities and reducing analytical drift), and chromatogram alignment. Following this, data preparation includes data integrity checking, data normalization, and compound name identification using the univariate, multivariate, clustering, and classification statistical analyses [32]. Following data processing, data interpretation, and metabolite identification from mass spectrum can be performed with the following: functional interpretation, enrichment analysis, pathway analysis, and metabolite pathway networks mapping. Commonly used tools include XCMS [33], Metaboanalyst [34], Progenesis [35], MetaCore [36], and 3Omics [37], with different analysis capabilities. The software processes raw mass spectrum data, perform various statistical analyses to find significantly altered ions/features, and for metabolite identifications connect to the metabolite database search such as Human Metabolome Database (HMDB) [14], Metabolite and Tandem MS Database (METLIN) [38], LIPID MAPS [39], Madison Metabolomics Consortium Database (MMCD) [40], BiGG [41], SetupX [42], KNApSAcK [43], and MetaboLights [44]. Compound or compound-specific databases include PubChem [45], Chemical Entities of Biological Interest (ChEBI) [46], ChemSpider [47], The KEGG GLYCAN [48], KEGG COMPOUND [49], and In Vivo/In Silico Metabolites Database (IIMDB) [50]. Metabolic pathway databases include Kyoto Encyclopedia of Genes and Genomes (KEGG) [51], BiKEGG [52], KEGG PATHWAY [53], MetaCyc [54], BioCyc [54], Model SEED [55], Reactome [56], and Ingenuity Pathway Analysis (IPA) [57].

3. Challenges and affecting factors

Though there are extensive tools and databases for analysis and identification of metabolites, challenges remain in the field of data analysis/integration, pathway analysis, and metabolite identification in untargeted metabolomics due to the high-throughput heterogeneous omics data which essentially requires improved bioinformatics and computational techniques to comprehensively evaluate the metabolomic profiles and completion of the human metabolome [32, 58–62].

Chromatographic resolution, absolute MS signals, and compound identification can be affected during ionization process due to polarity, ion sources, ion suppression, flow rates, and MS vacuum, sample preparation strategies, purity and temperature (prechilled) of reagents and solvents, different laboratory staff and techniques in different analysis days (typical sources of variance), different mass analyzers, chromatographic separation columns and compositions that cause forming of different fragments of the same molecule, frequent detection of the most abundant molecules, ion suppression, and day-day variation. Another challenge is the existence of isomers with identical masses and highly similar spectra which complicates distinguishing and differentiation during metabolite assignment to spectrum features [63, 64].

During the analytical process, a plot of the area internal standard (IS) response ratio vs. analyte concentration of each sample within a batch is performed to obtain calibration curves, which are needed to control for accuracy and reproducibility of the system. Performance monitoring in terms of plots of absolute MS signals, retention time points, and mass patterns/chromatographic peak shape of the analytes/ISs are required for maintaining sensitivity, integrity, and robustness of the analytical results. It has been reported that high system pressure and short columns with small particle sizes (<2 μm) lead to better signal:noise ratios than columns with larger particle sizes. Small particle-sized columns also affect the fastness of screening feature of MS [65–68].

Quality control (QC) sample plots of analytical parameters vs. retention times and peak shapes during calibration, sample preparation, and the analysis belonging to each batch as a quality check are essential for optimum standardization. All QC samples of each batch collected from each sample to create the pooled QC should be identical to each other and placed 1 in 5–1 in 10 samples in the analyses [16]. During sample preparation, a general contamination due to matrix compounds occurs and degrades the analytical system. In order to eradicate those possibilities, monitorization and randomization using technical QC and pooled QC samples which involve small aliquots of each biological sample, batch samples, and blanks considering peak shapes and retention times should be performed. Using a set of ISs, control samples, and blanks, performance of the method and monitorization have to be performed. In addition, randomization has to be performed on both sample preparation and the analysis order of system.

In order to eliminate or minimize methodological challenges, systematic errors, and bias factor due to analytical drift such as batch effect and obtain analytical sensitivity and specificity, reproducibility, accurate quantitation, and high recovery + low metabolite losses, each analysis should be carried out specifically under conditions of method optimization and optimum performance monitoring. Furthermore, using quality control samples and ISs during sample preparation, injection, extraction, fractionation, separation, detection, and normalization periods is essential. Generally, stable isotopes are used as the most identical ISs to compounds of interest in order to obtain accurate identification and to eliminate metabolite losses and ion suppression.

DOI: http://dx.doi.org/10.5772/intechopen.88563

4. Metabolomics in health and disease

Metabolomics has several applications in health and disease including precision/personalized medicine, single cell, epidemiologic population studies, metabolic phenotyping and metabolome-wide association studies (MWAS), precision metabolomics, and in combination with other omics disciplines as integrative omics [69]. Single-cell metabolomics and single-cell lipidomics technologies allow high-dimensional characterizations of individual cells, disease heterogeneity and complexity, identification, expression and abundance of disease-associated small molecules, metabolites, and by LC-MS/MS, GC-MS/MS, and live single-cell mass spectrometry (LSCMS) [70]. Imaging MS analysis of human breast cancer samples at the single-cell level revealed cell-cell interactions and tumor heterogeneity [71].

Clinical biomarkers and different metabotypes of disease severity correlated to exposures [72], and biological outcomes [73] have been studied and identified through metabolomics profiling, MWAS, and metabolomics fingerprinting and footprinting techniques in individuals and populations which will enable precision medicine and public healthcare [74–79]. Studies moving from genome-wide association studies (GWAS) to metabolome-wide association studies (MWAS) were first described in 2008 as "environmental and genomic influences to investigate the connections between phenotype variation and disease risk factors" [78–80]. Rattray and colleagues suggested exposotypes on single individual phenotypes and populations, in epidemiologic research, and disease risk using metabolome-wide association studies and impacts on precision medicine [79].

In conclusion, MS-based metabolomics/lipidomics provides a useful approach for both identification of disease-related metabolites in biofluids or tissue and also encompasses classification and/or characterization of disease or treatment-associated molecular patterns generated from metabolites [81, 82].

Author details

Sinem Nalbantoglu
Molecular Oncology Laboratory, Gene Engineering and Biotechnology Institute, TUBITAK Marmara Research Center, Kocaeli, Turkey

*Address all correspondence to: nalbantoglusinem@gmail.com

References

[1] Tsoukalas D, Alegakis A, Fragkiadaki P, Papakonstantinou E, Nikitovic D, Karataraki A, et al. Application of metabolomics: Focus on the quantification of organic acids in healthy adults. International Journal of Molecular Medicine. 2017;**40**(1):112-120

[2] Sun J, Beger DR, Schnackenberg KL. Metabolomics as a tool for personalizing medicine: 2012 update. Personalized Medicine. 2013;**10**:149-161. DOI: 10.2217/pme.13.8

[3] McKillop AM, Flatt PR. Emerging applications of metabolomic and genomic profiling in diabetic clinical medicine. Diabetes Care. 2011;**34**: 2624-2630. DOI: 10.2337/dc11-0837

[4] Guijas C, Montenegro-Burke JR, Warth B, Spilker ME, Siuzdak G. Metabolomics activity screening for identifying metabolites that modulate phenotype. Nature Biotechnology. 2018; **36**(4):316-320. DOI: 10.1038/nbt.4101

[5] Astarita G, Langridge J. An emerging role for metabolomics in nutrition science. Journal of Nutrigenetics and Nutrigenomics. 2013;**6**:181-200. DOI: 10.1159/000354403

[6] Baraldi E, Carraro S, Giordano G, Reniero F, Perilongo G, Zacchello F. Metabolomics: Moving towards personalized medicine. Italian Journal of Pediatrics. 2009;**35**:30. DOI: 10.1186/1824-7288-35-30

[7] Vander Heiden MG. Targeting cancer metabolism: A therapeutic window opens. Nature Reviews. Drug Discovery. 2011;**10**(9):671-684

[8] Gowda GA, Zhang S, Gu H, Asiago V, Shanaiah N, Raftery D. Metabolomics-based methods for early disease diagnostics. Expert Review of Molecular Diagnostics. 2008;**8**(5): 617-633

[9] Wang JH, Byun J, Pennathur S. Analytical approaches to metabolomics and applications to systems biology. Seminars in Nephrology. 2010;**30**(5): 500-511. DOI: 10.1016/j.semnephrol.2010.07.007

[10] Pauling L, Robinson AB, Teranishi R, Cary P. Quantitative analysis of urine vapor and breath by gas-liquid partition chromatography. Proceedings of the National Academy of Sciences of the United States of America. 1971;**68**(10):2374-2376

[11] Horning EC, Horning MG. Metabolic profiles: Gas-phase methods for analysis of metabolites. Clinical Chemistry. 1971;**17**(8):802-809

[12] Oliver SG, Winson MK, Kell DB, Baganz F. Systematic functional analysis of the yeast genome. Trends in Biotechnology. 1998;**16**(9):373-378

[13] Nicholson JK, Lindon JC, Holmes E. 'Metabonomics': Understanding the metabolic responses of living systems to pathophysiological stimuli via multivariate statistical analysis of biological NMR spectroscopic data. Xenobiotica. 1999;**29**(11):1181-1189

[14] Wishart DS, Knox C, Guo AC, et al. HMDB: A knowledgebase for the human metabolome. Nucleic Acids Research. 2009;**37**(Database issue): D603-D610

[15] Wishart DS, Feunang YD, Marcu A, Guo AC, Liang K, Vázquez-Fresno R, et al. HMDB 4.0: The human metabolome database for 2018. Nucleic Acids Research. 2018;**46**(D1):D608-D617

[16] Hyötyläinen T, Wiedmer S, editors. Chromatographic Methods in Metabolomics. United Kingdom: The Royal Society of Chemistry; 2013. DOI: 10.1039/9781849737272

[17] Marshall DD, Powers R. Beyond the paradigm: Combining mass spectrometry and nuclear magnetic resonance for metabolomics. Progress in Nuclear Magnetic Resonance Spectroscopy. 2017;**100**:1-16. DOI: 10.1016/j.pnmrs.2017.01.001

[18] Wishart DS. Metabolomics: The principles and potential applications to transplantation. American Journal of Transplantation. 2005;**5**(12):2814-2820

[19] Masson P, Alves AC, Ebbels TM, Nicholson JK, Want EJ. Optimization and evaluation of metabolite extraction protocols for untargeted metabolic profiling of liver samples by UPLC-MS. Analytical Chemistry. 2010;**82**(18): 7779-7786

[20] Lin L, Yu Q, Yan X, Hang W, Zheng J, Xing J, et al. Direct infusion mass spectrometry or liquid chromatography mass spectrometry for human metabonomics? A serum metabonomic study of kidney cancer. The Analyst. 2010;**135**:2970-2978

[21] Vermeersch KA, Styczynski MP. Applications of metabolomics in cancer research. Journal of Carcinogenesis. 2013;**12**:9

[22] Zhou M, Guan W, Walker LD, Mezencev R, Benigno BB, Gray A, et al. Rapid mass spectrometric metabolic profiling of blood sera detects ovarian cancer with high accuracy. Cancer Epidemiology, Biomarkers & Prevention. 2010;**19**:2262-2271

[23] Tolstikov VV, Fiehn O. Analysis of highly polar compounds of plant origin: Combination of hydrophilic interaction chromatography and electrospray ion trap mass spectrometry. Analytical Biochemistry. 2002;**301**(2):298-307. DOI: 10.1006/abio.2001.5513

[24] Rodrigues D, Pinto J, Araújo AM, Monteiro-Reis S, Jerónimo C, Henrique R, et al. Volatile metabolomic signature of bladder cancer cell lines based on gas chromatography-mass spectrometry. Metabolomics. 2018; **14**(5):62. DOI: 10.1007/s11306-018-1361-9

[25] Hofmann A, Clokie S, editor. In: Wilson and Walker's Principles and Techniques of Biochemistry and Molecular Biology. 8th edition. Cambridge University Press; 2018. 956 p. ISBN-10: 1107162270

[26] Kebarle P. A brief overview of the present status of the mechanisms involved in electrospray mass spectrometry. Journal of Mass Spectrometry. 2000;**35**:804-817

[27] Edwards JL, Kennedy RT. Metabolomic analysis of eukaryotic tissue and prokaryotes using negative mode MALDI time-of-flight mass spectrometry. Analytical Chemistry. 2005;**77**(7):2201-2209

[28] Pitt JJ. Principles and applications of liquid chromatography-mass spectrometry in clinical biochemistry. Clinical Biochemist Reviews. 2009; **30**(1):19-34

[29] Rosenberg E. The potential of organic (electrospray- and atmospheric pressure chemical ionisation) mass spectrometric techniques coupled to liquid-phase separation for speciation analysis. Journal of Chromatography. A. 2003;**1000**:841-889

[30] Byrdwell WC. Atmospheric pressure chemical ionization mass spectrometry for analysis of lipids. Lipids. 2001;**36**:327-346

[31] Dally JE, Gorniak J, Bowie R, Bentzley CM. Quantitation of underivatized free amino acids in mammalian cell culture media using matrix assisted laser desorption ionization time-of-flight mass

spectrometry. Analytical Chemistry. 2003;**75**(19):5046-5053

[32] Cambiaghi A, Ferrario M, Masseroli M. Analysis of metabolomic data: Tools, current strategies and future challenges for omics data integration. Briefings in Bioinformatics. 2017;**18**(3): 498-510

[33] Smith CA, Want EJ, O'Maille G, et al. XCMS: Processing mass spectrometry data for metabolite profiling using nonlinear peak alignment, matching, and identification. Analytical Chemistry. 2006;**78**:779-787

[34] Xia J, Psychogios N, Young N, et al. MetaboAnalyst: A web server for metabolomic data analysis and interpretation. Nucleic Acids Research. 2009;**37**:652-660

[35] Progenesis QI. Available from: h ttp://www.nonlinear.com/progenesis/ qi/ [Accessed: 25 February 2016]

[36] Thomson Reuters. MetaCoreTM 2004. Available from: http://lsresearch. thom sonreuters.com/ [Accessed: 25 February 2016]

[37] Kuo T-C, Tian T-F, Tseng YJ. 3Omics: A web-based systems biology tool for analysis, integration and visualization of human transcriptomic, proteomic and metabolomic data. BMC Systems Biology. 2013;**7**:64

[38] Smith CA, O'Maille G, Want EJ, et al. METLIN: A metabolite mass spectral database. Therapeutic Drug Monitoring. 2005;**27**:747-751

[39] Fahy E, Sud M, Cotter D, Subramaniam S. LIPID MAPS online tools for lipid research. Nucleic Acids Research. 2007;**35**(Web Server issue): W606-W612. DOI: 10.1093/nar/gkm324

[40] Cui Q, Lewis IA, Hegeman AD, Anderson ME, Li J, Schulte CF, et al. Metabolite identification via the Madison Metabolomics Consortium Database. Nature Biotechnology. 2008; **26**:162-164

[41] King ZA, Lu J, Dräger A, Miller P, Federowicz S, Lerman JA, et al. BiGG models: A platform for integrating, standardizing and sharing genome-scale models. Nucleic Acids Research. 2016; **44**(D1):D515-D522. DOI: 10.1093/nar/ gkv1049

[42] Scholz M, Fiehn O. SetupX—A public study design database for metabolomic projects. Pacific Symposium on Biocomputing. 2007: 169-180

[43] KNApSAcK: A Comprehensive Species-Metabolite Relationship Database. Available from: http://kanaya .aist-nara.ac.jp/KNApSAcK/

[44] Kale NS, Haug K, Conesa P, Jayseelan K, Moreno P, Rocca-Serra P, et al. MetaboLights: An open-access database repository for metabolomics data. Current Protocols in Bioinformatics. 2016;**53**:14.13.1-14.1318. DOI: 10.1002/ 0471250953.bi1413s53

[45] https://pubchem.ncbi.nlm.nih.gov/

[46] Hastings J, Owen G, Dekker A, Ennis M, Kale N, Muthukrishnan V, et al. ChEBI in 2016: Improved services and an expanding collection of metabolites. Nucleic Acids Research. 2016;**44**(D1):D1214-D1219. DOI: 10.1093/nar/gkv1031

[47] https://www.chemspider.com/Data sourceDetails.aspx?id=84

[48] https://www.genome.jp/kegg/ glycan/

[49] https://www.genome.jp/kegg/c ompound/

[50] Menikarachchi LC, Hill DW, Hamdalla MA, Mandoiu II, Grant DF. In silico enzymatic synthesis of a 400,000

compound biochemical database for nontargeted metabolomics. Journal of Chemical Information and Modeling. 2013;**53**(9):2483-2492. DOI: 10.1021/ci400368v

[51] https://www.genome.jp/kegg/

[52] Jamialahmadi O, Motamedian E, Hashemi-Najafabadi S. BiKEGG: A COBRA toolbox extension for bridging the BiGG and KEGG databases. Molecular BioSystems. 2016;**12**(11): 3459-3466. DOI: 10.1039/c6mb00532b

[53] https://www.genome.jp/kegg-bin/show_pathway?map01100

[54] Caspi R, Altman T, Dreher K, Fulcher CA, Subhraveti P, Keseler IM, et al. The MetaCyc database of metabolic pathways and enzymes and the BioCyc collection of pathway/genome databases. Nucleic Acids Research. 2012;**40**(Database issue): D742-D753. DOI: 10.1093/nar/gkr1014

[55] Devoid S, Overbeek R, DeJongh M, Vonstein V, Best AA, Henry C. Automated genome annotation and metabolic model reconstruction in the SEED and model SEED. Methods in Molecular Biology. 2013;**985**:17-45. DOI: 10.1007/978-1-62703-299-5_2

[56] Jupe S, Akkerman JW, Soranzo N, Ouwehand WH. Reactome—A curated knowledgebase of biological pathways: Megakaryocytes and platelets. Journal of Thrombosis and Haemostasis. 2012; **10**(11):2399-2402. DOI: 10.1111/j.1538-7836.2012.04930.x

[57] Ingenuity. IPA: Ingenuity Pathway Analysis. Available from: http://www.ingenuity.com/products/ipa/ [Accessed: 25 February 2016]

[58] Nanda T, Das M, Tripathy K, Ravi Teja Y. Metabolomics: The future of systems biology. Journal of Computer Science and Systems Biology. 2011:S13. DOI: 10.4172/jcsb.S13-003

[59] Taylor PJ. Matrix effects: The Achilles heel of quantitative high-performance liquid chromatography-electrospray-tandem mass spectrometry. Clinical Biochemistry. 2005;**38**:328-334

[60] Roberts LD, Souza LA, Gerszten RE, et al. Targeted metabolomics. Current Protocols in Molecular Biology. 2012;**30**:2

[61] Annesley TM. Ion suppression in mass spectrometry. Clinical Chemistry. 2003;**49**:1041-1044

[62] Matuszewski BK, Constanzer ML, Chavez-Eng CM. Strategies for the assessment of matrix effect in quantitative bioanalytical methods based on HPLC-MS/MS. Analytical Chemistry. 2003;**75**:3019-3030

[63] Oglesbee D, Sanders KA, Lacey JM, Magera MJ, Casetta B, Strauss KA, et al. Second-tier test for quantification of alloisoleucine and branched-chain amino acids in dried blood spots to improve newborn screening for maple syrup urine disease. Clinical Chemistry. 2008;**54**:542-549

[64] Pitt JJ, Eggington M, Kahler SG. Comprehensive screening of urine samples for inborn errors of metabolism by electrospray tandem mass spectrometry. Clinical Chemistry. 2002; **48**:1970-1980

[65] Nordstrom A, O'Maille G, Qin C, Siuzdak G. Nonlinear data alignment for UPLC-MS and HPLC-MS based metabolomics: Quantitative analysis of endogenous and exogenous metabolites in human serum. Analytical Chemistry. 2006;**78**:3289-3295

[66] Aronov PA, Hall LM, Dettmer K, Stephensen CB, Hammock BD. Metabolic profiling of major vitamin D metabolites using Diels-Alder derivatization and ultra-performance liquid chromatography-tandem mass

spectrometry. Analytical and Bioanalytical Chemistry. 2008;**391**: 1917-1930

[67] Ventura R, Roig M, Montfort N, Saez P, Berges R, Segura J. High-throughput and sensitive screening by ultra-performance liquid chromatography tandem mass spectrometry of diuretics and other doping agents. European Journal of Mass Spectrometry (Chichester, England). 2008;**14**:191-200

[68] Licea-Perez H, Wang S, Szapacs ME, Yang E. Development of a highly sensitive and selective UPLC/MS/MS method for the simultaneous determination of testosterone and 5alpha-dihydrotestosterone in human serum to support testosterone replacement therapy for hypogonadism. Steroids. 2008;**73**:601-610

[69] Khoomrung S, Wanichthanarak K, Nookaew I, Thamsermsang O, Seubnooch P, Laohapand T, et al. Metabolomics and integrative omics for the development of Thai traditional medicine. Frontiers in Pharmacology. 2017;**8**:474. DOI: 10.3389/fphar.2017.00474

[70] Emara S, Amer S, Ali A, Abouleila Y, Oga A, Masujima T. Single-cell metabolomics. Advances in Experimental Medicine and Biology. 2017;**965**:323-343. DOI: 10.1007/978-3-319-47656-8_13

[71] Giesen C, Wang HA, Schapiro D, Zivanovic N, Jacobs A, Hattendorf B, et al. Highly multiplexed imaging of tumor tissues with subcellular resolution by mass cytometry. Nature Methods. 2014;**11**(4):417-422. DOI: 10.1038/nmeth.2869

[72] Walker DI, Pennell KD, Uppal K, Xia X, Hopke PK, Utell MJ, et al. Pilot metabolome-wide association study of benzo(a)pyrene in serum from military personnel. Journal of Occupational and Environmental Medicine. 2016;**58**: S44-S52

[73] Bictash M, Ebbels TM, Chan Q, Loo RL, Yap IKS, Brown IJ, et al. Opening up the "black box": Metabolic phenotyping and metabolome-wide association studies in epidemiology. Journal of Clinical Epidemiology. 2010; **63**:970-979

[74] Holmes E, Nicholson JK. Human metabolic phenotyping and metabolome wide association studies. Ernst Schering Foundation Symposium Proceedings. 2007;**4**:227-249

[75] Holmes E, Wilson ID, Nicholson JK. Metabolic phenotyping in health and disease. Cell. 2008;**134**(5):714-717. DOI: 10.1016/j.cell.2008.08.026

[76] Nicholson JK, Holmes E, Elliott P. The metabolome-wide association study: A new look at human disease risk factors. Journal of Proteome Research. 2008;**7**:3637-3638

[77] Tolstikov V. Metabolomics: Bridging the gap between pharmaceutical development and population health. Metabolites. 2016;**6**(3). DOI: 10.3390/metabo6030020.

[78] Beger RD, Dunn W, Schmidt MA, Gross SS, Kirwan JA, Cascante M, et al. Metabolomics enables precision medicine: "A White Paper, Community Perspective". Metabolomics. 2016; **12**(10):149. DOI: 10.1007/s11306-016-1094-6

[79] Chadeau-Hyam M, Ebbels TM, Brown IJ, Chan Q, Stamler J, et al. Metabolic profiling and the metabolome-wide association study: Significance level for biomarker identification. Journal of Proteome Research. 2010;**9**:4620-4627

[80] Rattray NJW, Deziel NC, Wallach JD, Khan SA, Vasiliou V, Ioannidis JPA, et al. Beyond genomics:

Understanding expósotypes through metabolomics. Human Genomics. 2018; **12**(1):4. DOI: 10.1186/s40246-018-0134-x

[81] Liu R, Zhang G, Sun M, Pan X, Yang Z. Integrating a generalized data analysis workflow with the single-probe mass spectrometry experiment for single cell metabolomics. Analytica Chimica Acta. 2019;**1064**:71-79. DOI: 10.1016/j.aca.2019.03.006

[82] Yang K, Han X. Lipidomics: Techniques, applications, and outcomes related to biomedical sciences. Trends in Biochemical Sciences. 2016;**41**(11): 954-969. DOI: 10.1016/j.tibs.2016.08.010